TACKLING THE IMAGE CLASSIFICATION PROBLEM WITH SPIKING NEURAL NETWORK TECHNIQUE

JOSHUA ARUL KUMAR R

TACKLING THE IMAGE CLASSIFICATION PROBLEM WITH SPIKING NEURAL NETWORK TECHNIQUE

First Edition September 2024

Written by JOSHUA ARUL KUMAR R

TABLE OF CONTENTS

LIST OF TABLES

LIST OF SYMBOLS AND ABBREVIATIONS

ADAM	Adaptive Moment Estimation optimizer
ANN	Artificial Neural Network
ASIC	Application Specific Integrated Circuit
b_i	Bias for class i of softmax
CNN	Convolutional Neural Network
CPU	Central Processing Unit
C	The membrane capacitance value
DL	Deep learning
$\frac{d}{d(t)}$	differentiation operator
$evidence_i$	Evidence for a class i of softmax
$\exp i$	exponentiation operator
FLOPS	Floating Point Operations Per Second
FPGA	Field Programmable Gate Array
F	Spatial extent of CNN
GMDH	Group Method of Data Handling
GPU	Graphics Processing Unit
IF	Integrate-and-Fire
j	Constant Input Current
$J(t)$	Input current to the neuron over time
K	Number of filters in a layer in CNN
LIF	Leaky Integrate-and-Fire
LMS	Least Mean Square
ML	Machine Learning
MLP	Multi-layer Perceptron
MNIST	Modified National Institute of Standards and Technology
mrFPGA	memristor-based reconfiguration FPGA architecture

NEF	Neural Engineering Framework
P	Amount of zero padding in CNN
Q	Charge across a capacitor
r	The spike rate
R	The membrane resistance
RELU	Rectified Linear Unit
RMSProp	Root Mean Square Propagation optimizer
RNN	Recurrent Neural Network
S	Stride of a CNN
SGD	Stochastic Gradient Descent
SN-CNN	Spiking Neuron - Convolution Neural Network
SNN	Spiking Neural Networks
STDP	Spike-Timing Dependant Plasticity
SVC	Support Vector Classification
Σ	summation operator
τ_{rc}	Time Constant
TPU	Tensor Processing Unit
t_{ref}	Time in resting potential
V	Voltage across the capacitor
V_{th}	Neuron threshold voltage
V_{rest}	Resting potential of membrane
W	Size of the input volume in CNN
W_i	Weights of softmax
y	Predicted probabilities in Softmax

CHAPTER 1

NEED FOR BIOLOGICALLY INSPIRED OBJECT RECOGNITION

This chapter explains the need for better image classification models and how a bio-plausible data-driven approach opens an opportunity to solve the problem effectively.

1.1 INTRODUCTION

Vision inference is the process of the eye, recognizing an object, analyzing the same and reconstructing a complete picture of an object. Unlike some mental processes like computing an addition, or reasoning for a decision in a chess game, vision inference is a mental ability that the person involved is not consciously aware of. As a result, many people including experienced researchers often underestimate the difficulty of recognizing the image while modeling the same. The earliest pattern recognition research dates back to 1966 when Marvin Minsky and Seymour Papert tried developing an object recognition system (Papert 1966). Yet it has taken more than 50 years to develop a system for image classification which is similar in accuracy to an average person. In the last decade, machine learning has succeeded in tasks like identifying a wide variety of objects and faces in realistic situations and contexts.

The key to success is the focus of shift to object identification, based on the developments observed in the field of neuroscience. Since the brain of humans and other animals are the only systems that solve these difficult problems, they are the main source of inspiration on how to solve image classification problems with machines.

1.1.1 Paradigm Shift in Computing Systems

On the other hand, computing technology is evolving in different dimensions due to the technological advancements and saturation that are encountered by the existing systems. There is a paradigm shift in the use of computing devices, for example from computers to handheld devices to wearables. Low power consumption has become the key for design, as the applications moved from computers to communication devices and sensors in recent years. This led to the search for different alternatives in place of the Von Neumann architecture. Moore's law, the primary driver of integrated circuit technology has also taken a backseat in recent years (Calimera et al. 2013). The primary reason is the alterations in the way computing is looked over in terms of its performance. Moore's law addresses the increase in integration density by device size reduction, increasing chip area and device and circuit cleverness (Peper 2017), all of which are dependent on technology.

Nowadays, functionality has priority over clock speed or device feature size, with more focus on performance improvement rather than technology. There is a sudden shift from the GHz war between manufacturers since 2004, as Dennard's Scaling rules became unsustainable. It also had a paradigm shift from looking at computations with FLOPS as the benchmark to FLOPS per watt emphasizing energy efficiency. The merging of logic and memory is attributed as a solution, to bring the processing of information closer to where it is stored, thereby addressing the Von Neumann's bottleneck as well as providing power efficiency.

As a result, researchers started shifting their attention to brain-inspired computing. Neurons form a complex network in the brain that comes with remarkable abilities. If Moore's Law is extrapolated, the number of neurons in the human brain (100 billion) will be equivalent to the number of transistors in a chip only by 2026 (Henderson 2017), although this has nothing in comparison to the performance of the human brain.

1.1.2 Bio-inspired Computing Systems

Cellular Automata (CA) with its high degree of fine-grained parallelism evolved as a new model. It is extremely regular, brain-inspired computing and tends to rely more on random connectivity. CAs are Delay-Insensitive (DI) and its functionality is determined by its transition rules. Cell complexity depends on the number of states a cell can assume and the number of transition rules. However, CA could not emerge as a promising solution owing to the following challenges. Low functional density makes it necessary to store transition rules in each cell leading to increased overhead. It is hard to find materials in nature that implement CA. Finally, there is a requirement of energy to be fed in order to sustain computation.

There exists a vast asymmetry between the working pattern of the brain and traditional computers. While the later can play with numbers at an unprecedented speed, is reliable, not fatigued and unbiased, human brain on the other side can think "out of the box", detect patterns at a faster rate, and is not necessary to be programmed often, as it is good at learning (Al-Rodhan 2017). All these can be done with a power of fewer than 20 watts. Creativity and imagination cannot be accomplished with traditional computers. These limitations faced by traditional computing led to the exploration in the field of Neuromorphic computing.

1.2 MOTIVATION OF RESEARCH

The main goal of an Image Classification system is to process data into meaningful information. The processed information during classification is helpful to categorize images into various groups. A good image classification system depends on the determination of proper classification system, right and required feature extraction, choice of good training samples, suitable selection of classification method, image pre and post processing and better assessment of overall accuracy (Lu and Weng 2007). The solution to classification problems can be carried out mathematically and in a nonlinear fashion. The task is challenging and depends on the accuracy and distribution of data properties as well as the capabilities of the implemented models.

A plethora of classifiers are available and can be categorized on the basis of a) characteristic used as shape-based and motion-based, b) training sample used as supervised and unsupervised, c) assumption of parameters as parametric and non-parametric, d) pixel information usage as per-pixel, sub-pixel, per-field and object-oriented, e) spatial information as spectral and contextual, and f) number of outputs for each spatial element as hard classification and soft classification (Kamavisdar et al. 2013).

Machine learning (ML) algorithms helped to make machines do intelligent things that were more focused on developing rule-based systems using artificial neural networks (ANN) or Fuzzy logic. But it is difficult to come up with the best rules covering a majority of image classification situations and even such rules are fragile to changes in viewing angle. This led to shift the focus on a data-driven approach where more generalization is possible. Hence convolution neural network (CNN) gained importance.

However, when implementation in low commodity hardware is

concerned, it is still a real challenge. With the paradigm shift in usage of devices from computers to mobiles classification, low power and less computational devices become a necessity.

1.3 STATEMENT OF RESEARCH

Spiking neurons (SN) use dynamic event-driven processing and has a strong and fast adaptation capability. They outperform the traditional neurons in terms of complexity, calculability, and regularization (Paugam-Moisy et al. 2002). While deep learning (DL) using CNN addresses the large scale learning problem, the use of SN adds more bio-plausibility to the approach while using a smaller chip area. A combination of these two methodologies provides the best choice of a neuromorphic system for classification problems.

Hence, this research aims at combining machine learning, with concepts of computational neuroscience to create a model for Image classification and is known as the SN-CNN model. The model is biologically inspired, data-driven and realistic in scaling to high-level implementations by using the spiking neural network (SNN). The architecture was implemented and tested in a core i5 processor. The results obtained were in par with CNN while the advantage is observed in the quick arrival of results and ease of implementation in low commodity hardware and usage of less chip area.

1.4 ORGANIZATION OF THE THESIS

The thesis is organized into the following Chapters.

- Chapter 1 mentions the necessity for developing bio-plausible approach for image classification. It mentions the motivation behind

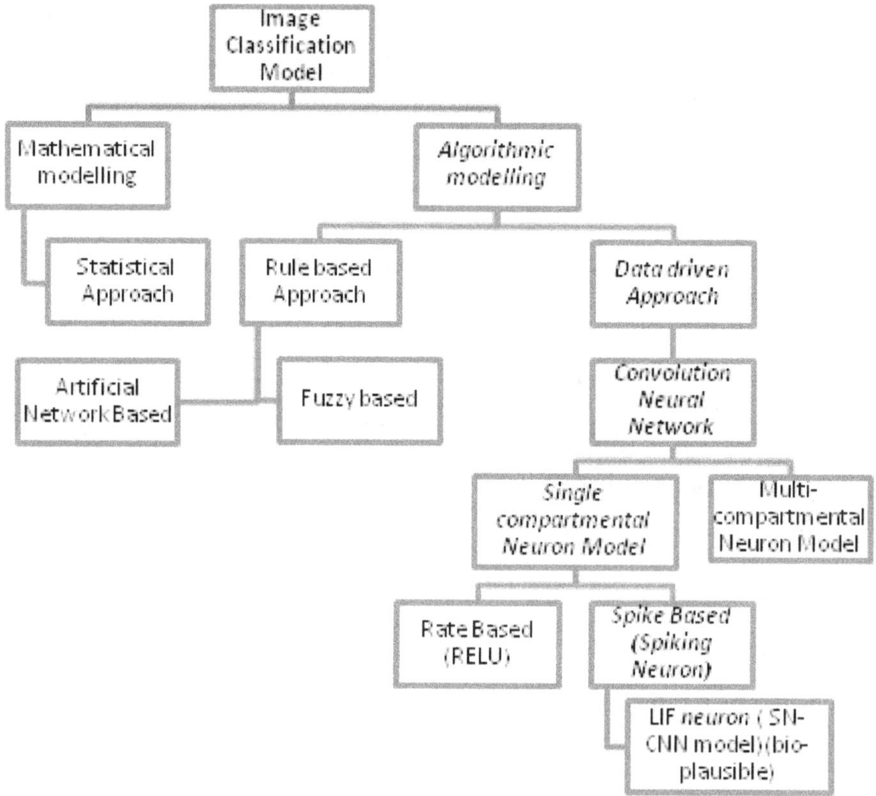

Figure 1.1: Flow diagram of SN-CNN model

this research, the statement of research followed by the thesis organization.

- Chapter 2 provides a review of literature in the fields of machine learning and computational neuroscience in lieu of image classification. It also discusses the avenues for combining both of them so as to obtain better image classification models.

- Chapter 3 discusses about the developments in deep learning towards image classification. It also provides a short insight on integrating deep learning with neuromorphic modelling. This is achieved by the use of bio-inspired neuron model. A data-driven approach for image classification using CNN and integration of spiking neuron to form SN-CNN model is also presented. A choice on the encoding information in a bio-plausible manner is analysed. The approach towards developing single neuron model on the basis of neural engineering framework (NEF) is presented which allows largescale implementations.

- Chapter 4 provides the basic components required for constructing a ML system. It is followed by setting the design objective for SN-CNN construction. The factors that led to the choice of Nengo, NengoDL simulator and Tensorflow for simulation of SN-CNN model is presented. Then the construction, choice of loss function, optimizer and the testing process for SN-CNN model is presented.

- Chapter 5 discusses the simulation results of SN-CNN model. It also demonstrates the performance of results obtained when the model is tried with differnt number of epoch, optimizers, architectural changes and neuron options. The best of the results are chosen towards formation of Improved SN-CNN model and is tested and evaluated in Fashion MNIST dataset. The empirical results are presented and the tuning process is explained. A comparison of results with conventional models of Fashion MNIST is also presented.

- Chapter 6 discusses the challenges of implementation of ANN models in hardware platforms. It also provides the opportunities of implementation of SN-CNN in FPGA and neuromorphic hardware and discusses the advantages and scope of such implementations.

- Conclusion summarises the key contributions of this research and provides the future scope in improving the bio-plausible, data-driven SN-CNN model.

CHAPTER 2

A REVIEW OF NEURAL NETWORK MODELS AND MACHINE LEARNING MECHANISM

2.1   COMPUTATIONAL NEUROSCIENCE

Computational neuroscience is a field related to studying individual neurons, their combinations in successive larger network in a meaningful way.   This is unlike psychology and cognitive science where the behaviour of creatures is studied first and then trying to understand the role of brain in these behaviours. Hence psychology and cognitive science holds a top-down approach whereas computational neuroscience takes a bottom-up approach in neuroscience.   Thus it takes a computational approach in understanding the performance of brain and the network of its elements.   Traditionally it aims at developing mathematical models of individual neurons, extending them to small and medium sized networks. However recently the fields of computational neuroscience and cognitive science began to meet which provided an opportunity for the development of neurally detailed models that reproduce organism-level behaviours. Hence it is necessary to look into the biological details as well as the concepts from computational neuroscience that is used in this thesis.

2.1.1   Neuron Physiology

Neurons are the main computation entities of the brain. They receive input from other neurons and on sufficient excitation, will fire an action potential

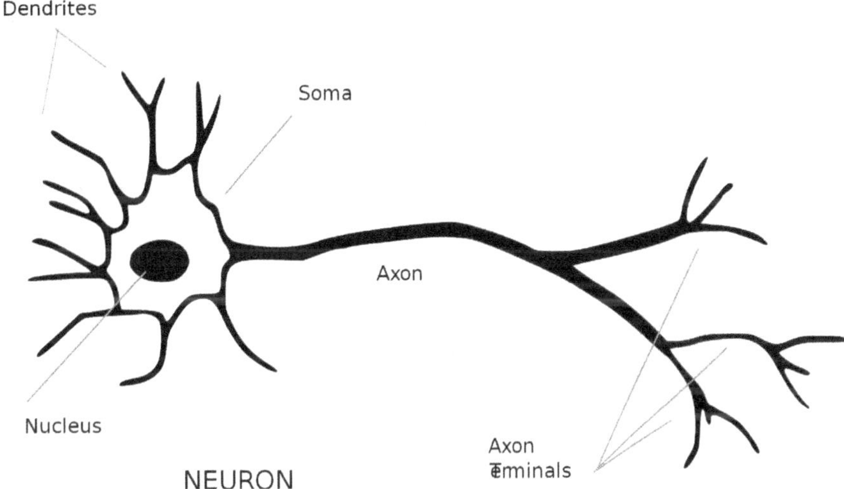

Figure 2.1: Basic structure of neuron

or spike which is propagated to other neurons. A basic structure of the Neuron is shown in Figure 2.1.

Each neuron is divided into three parts namely the dendrites, the soma and the axon. The dendrites receive input current and then transmit into the cell body called soma. When a neuron spikes, the current is traversed into axon causing neurotransmitters to release at the synapses. Synapse acts as a bridge between neuron's axon to the dendrites of other neurons. The released neurotransmitters from the axon results in dendrite input currents in its connected neurons. Soma is the main body of neuron where an action potential is created by integration of currents from dendrites. The soma has a negative charge when a neuron is at rest. This voltage called resting voltage is maintained by a particular concentration of ions (mostly sodium, potassium and calcium) inside the cell by ion pumps.

Depolarization of the cell begins as currents arrive from the dendrites. When the voltage at soma is high enough, it begins to trigger voltage-activated sodium channels thereby allowing sodium ions to enter the cell furthering depolarization. This process continues till the electrical gradient caused as a result of increase in sodium ions, oppose the chemical gradient due to imbalance of sodium inside and outside the cell. This large depolarisation creates a voltage spike as the neuron has much more positive charge than the resting voltage. It also triggers voltage gated potassium channels, so as to let the potassium ions out of the cell repolarising it, and bringing the cell to hyperpolarized state which is below the resting potential. During repolarization the potassium channels remain open and the sodium channels is inactivated making the neuron impossible to fire. This duration is known as absolute refractory period. The depolarization due to action potential increases the voltage potential at soma and also some depolarization in parts of axons near the soma. This in turn triggers sodium channels further down the axon. Thus the somatic spike triggers a voltage wave down the axon which enables the synaptic vesicles at the other end to release neurotransmitters.

Axons vary widely in length and are responsible for long range transmission of signals in the brain. Axons are coated in myelin composed mainly of lipids which act as a good electrical insulator. Dendrites, on the other hand are a branched structure that forms the main input pathway of a neuron. It sums the output current received from synapses of the surrounding neurons in the form of an electrical potential, which diffuses along the tree to the soma. Traditionally the signal summation is believed to be linear, but recent studies (Polsky et al. 2004) show that signal summation in dendrites is more complex and is a combination of linear and nonlinear (sigmoidal) elements.

2.1.1.1   Spiking dynamics of a neuron

Neurons cells are responsible for integrating, encoding and transmitting, signals originating inside or outside the nervous system. The input a neuron receives at synapses from other neurons, cause transient changes in its membrane potential, called post synaptic potential. The changes in the membrane potential in a neuron helps in transmission of information within and between neurons. These changes in potential are facilitated by the flux of ions through ion channels present in the membrane.

The ion channels open or close depending on the release of neurotransmitters which repolarize or depolarize the cell. When the postsynaptic potential reaches a threshold, the neuron produces an impulse or spike. These spikes also known as action potentials, are the units of information transmission at the inter neuronal level. Information is believed to be encoded in terms of timing of action potential (i.e., Temporal coding) as well as frequency of the action potentials called spiking or firing rate (i.e., rate coding) (Deco et al. 2008).

2.2   NEURAL NETWORK MODELS

There seems to be many options in the choice of neural network models. In some cases, a particular application area may drive the choice. A pattern classification problem may suggest a more biologically plausible model whereas a pattern identification task may require a computationally driven model, owing to its high accuracy requirement. In other cases, the characteristics of a particular device or material may dominate the choice. There seems to be a variety of models that range from those predominantly bio-inspired, to that of computation driven. However, a neural network model in general defines the choice of components to make up the network, their operations and the interactions between them.

The various neural network models can be grouped under three categories namely Neuron models, Synapse models and Network models as shown in Figure 2.2.

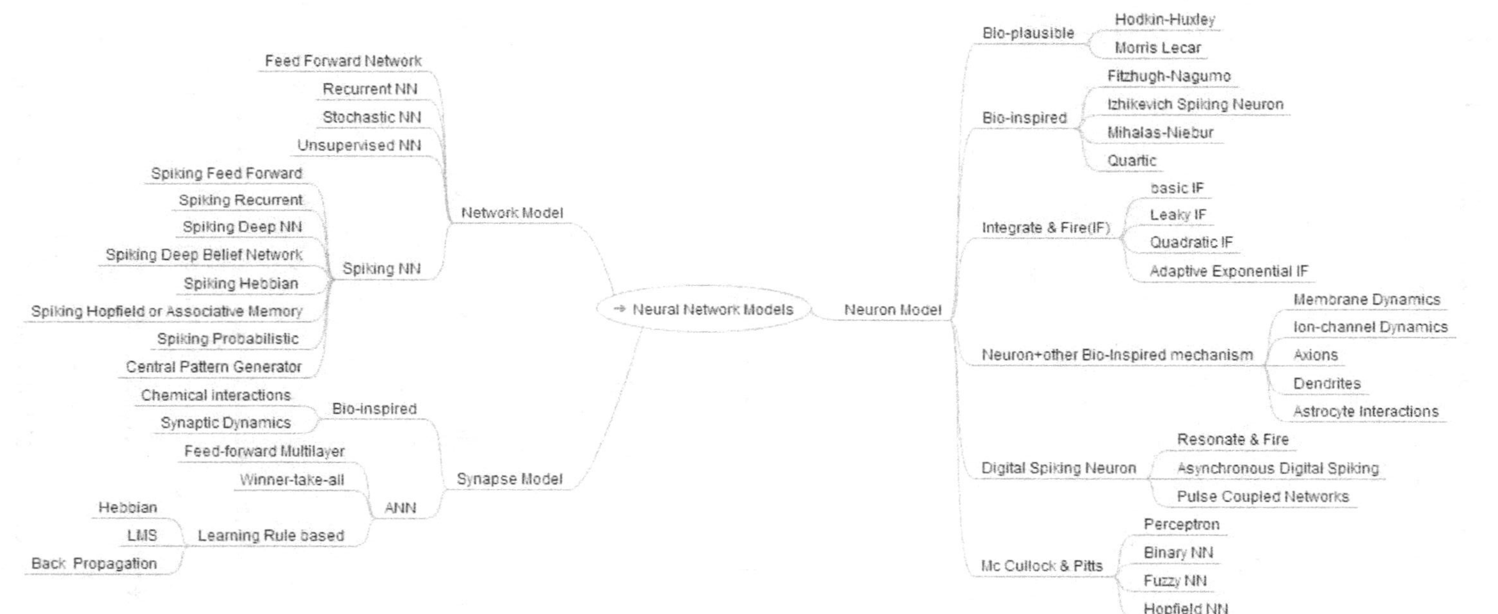

Figure 2.2: A mindmap of Neural Network Models

2.2.1   Neuron Models

The behaviour of a biological neuron is to accumulate charges received from various synapses attached to it through a change in voltage potential across the neuron's cell membrane. Hence, these models have some concept of charge accumulation and firing to affect other neurons. They can further be classified into:

2.2.1.1   Bio-plausible

Here the behaviour of the model is built as seen in biological neural system. The model's behaviour is an equivalent to biological system.

Hodgkin-Huxley model (Abbott and Kepler 1990) which uses a 4D nonlinear differential equation, tries to create an accurate bio-neural system. Here the implementation is carried out by using electric circuits made of resistors and capacitors. It models electro-chemical information transmission of natural neurons to some extent. This model involves a computational cost of 1200 floating point computations (FLOPS) per 1ms simulation.

Morris Lecar Model (Lecar 2007) on the other hand uses a simple 2D non-linear differential equation and involves a computational cost of one or several hundred FLOPS.

2.2.1.2   Biologically inspired

This category mimics the behaviour of biological neural system but not in the bio plausible way. Rather, it considers the computational method also.

- Fitzhugh-Nagumo (Izhikevich and FitzHugh 2006) and Hindmarsh-Rose (Hindmarsh and Cornelius 2005) models: These are simplified version of hodgkin-huxley model. They use simple computations and less number of parameters. It is easy to set and train as there are less parameters. It is simple to implement in a small footprint.

- Izhikevich spiking neuron model (Izhikevich 2003): This model is popular due to its simple computation as well as the ability to reproduce bio-accurate behaviour. They exhibit bursting and spiking behaviour and is a good compromise between biophysical plausibility and computational cost (13 FLOPS).

- Mihalas-Niebur Neuron model (Folowosele et al. 2011): This model also replicates bursting and spiking behaviour. It is defined by a set of linear differential equations and has neuromorphic implementations.

- Quartic model (Grassia et al. 2014): This model uses two non-linear differential equations and is implementable in the neuromorphic system.

These models are based on better trade-off between bio-plausibility and computation and involve differential equations.

2.2.1.3   Neuron model with additional bio-inspired mechanism

Few other models are developed which include modelling of neurons with other additional components like dendrites, axons etc. Such models have added components like membrane dynamics (Arthur and Boahen 2011), ion-channel dynamics (Hynna and Boahen 2006), and glial cell or astrocyte interactions (Hayati et al. 2016).

2.2.1.4   Integrate-and-fire

These form the simple category of bio-inspired spiking neuron model. They are less biologically realistic, but produce enough complexity in behaviour. The variety of models include:

- Basic integrate-and-fire (Abbott 1999a): This model treats biological neurons as point dynamical systems and the spatial structure is neglected here. It maintains current charge level of the neurons. The shape of the action potentials is neglected and every spike is considered as a uniform event defined only by the time of its appearance.

- Leaky integrate-and-fire (Tal and Schwartz 2006): This is one of the most popular model in neuromorphic system. It includes a leaky term to model, which causes the potential of the neuron to decay over time. It has the ability to reproduce neuro-computational properties and is limited to three different firing schemes that require only five FLOPS.

- Quadratic integrate-and-fire (Basham and Parent 2009): These are nonlinear integrate and fire models and require around 10 FLOPS.

- Adaptive exponential integrate-and-fire : It shares the features of Leaky integrate-and-fire and 2-D integrate and fire models with excellent spike time predictions (Abbas and Muthulakshmi 2014).

2.2.1.5   Digital spiking neuron

These models are characterised by cellular automata theory as opposed to a set of nonlinear or linear differential equations.

They include resonate and fire (Hishiki and Torikai 2009), rotate and fire (Hishiki and Torikai 2010), asynchronous digital spiking neuron model (Matsubara and Torikai 2011) and pulse coupled networks (Wang et al. 2010).

2.2.1.6   Mc'Cullock and Pitts model

These are the primitive models used in ANN. They are derivatives of Mc'Cullock and Pitts neuron. The most basic model is the perceptron model. This model is commonly used in hardware implementations. They have simple thresholding where the weights are modified based on activation functions. The most basic activation functions employed are sigmoid function and hyperbolic tangent function. Later, many hardware based activation functions like ramp-saturation function(Bañuelos-Saucedo et al. 2003), piecewise linear (Hikawa 2003), multi-threshold (Zhu et al. 2005), radial basis function, tangent sigmoid function (Sahin and Koyuncu 2012) and periodic activation function (Merkel et al. 2013) were involved. The other models which were developed from perceptron are binary neural network neurons (Deshmukh et al. 2005), fuzzy neural network neurons (Yamakawa 1996) and hopfield neural network neurons (Hollis and Paulos 1992).

2.2.2   Synapse Models

Modelling the dynamics of synapses is also important along with modeling internal neural dynamics. The simplest model of a synapse can be observed as a first-order lowpass filter. The impulse response describes the nature of postsynaptic current and the synaptic time constant is a direct proportion of the length of time the postsynaptic current spreads. Mainen et al. (Mainen and Sejnowski 1995) introduced a second-order lowpass filter better than the first-order lowpass filter model. Both are used as current-based models.

Synapse models are evolved for neuromorphic systems, since synapses are the most abundant elements occupying more chip area. Optimizing them is essential for better hardware implementation. Synapse models are relatively simple and its complexity increases with bio-plausibility. However, the compelling reason to build complex synapse models is its plasticity mechanism, which has been found to be related to learning in the biological brain and cause the strength of neurons to change over time.

Synapse models can be categorised as follows.

2.2.2.1   Biologically inspired synapse implementation

In Bio-inspired neuromorphic systems, synapse implementations model chemical interactions of synapse such as ion pumps or neuro transmitter interactions. Conductance based synapse models (Noack et al. 2014) implement ion channels also. Spiking neuromorphic systems like Spike-Timing Dependant Plasticity (STDP) exhibit plasticity and learning mechanism inspired by potentiation and depression. Similarly, model based on synaptic dynamics focus on shape of outgoing spike from synapse or post synaptic potential.

2.2.2.2   ANN synapse implementation

Non bio-inspired synapse models are also developed. Few of them are feed forward multi-layer networks (McGinnity et al. 1998), winner-take-all (Yuille and Grzywacz 1989) and convolutional neural network (Vianello et al. 2017). ANN synapse implementations are also based on learning rules like hebbian learning (Schneider and Card 1991) which by modifying synaptic strengths lead to the reorganization of connections within a neural network.

Other implementations based on learning includes least mean-square (Srinivasan et al. 2005) and back propagation learning rule (Rumelhart et al. 1988).

2.2.3 Factors Influencing the Choice of Neuron and Synapse Models

Neuron and synapse models for neuromorphic computing depends on the factors like computed dynamics, scaling, reproducibility, multiplexing, low power and intrinsic dynamics depending on the type of application. The existing neuron models mentioned provide a variety of tradeoff between biological inspiration, complexity and computational speed. Since they act as base entity, a careful choice of neuron model provides a better choice of creating a vibrant neuromorphic computing system.

The Synapse models choice is based on challenges in minimial use of chip area, as these act as memory and require more area when implemented in silicon. Also, the developmental issues in back end memory like memristors makes its implementation more difficult. However based on the amount of memory usage for different applications the synapse models can be selected.

2.2.4 Network Models

Network models primarily focus on the interconnections between different neurons and synapses and their way of interaction. A wide variety of network models are available for neuromorphic systems. The prime factors that govern the design of a network model are a) biological inspiration, b) topology of network, c) feasibility and applicability of training or learning algorithm and d) type of applications to be used.

2.2.4.1   Feed forward network model

The most popular network model in artificial neural networks are feed forward network models. The basic feed forward network is characterised by multilayer perceptrons. A special case where the number of weights after random assignment is never updated based on learning is extreme learning machines (Yao et al. 2013). Few more variants of feed forward networks include multilayer perceptrons with delay, probabilistic neural networks which involves Bayesian calculations (Aibe et al. 2002), single layer feed forward utilising radial basis functions as activation function and convolutional neural networks.

2.2.4.2   Recurrent neural network (RNN) model

These models allow for cycles in the network thereby exhibiting dynamic temporal behaviour. The various models developed under RNN are non-spiking RNN (Cauwenberghs 1994), Reservoir computing model where RNN is used as reservoir and its output is fed into simple feed forward network (Kudithipudi et al. 2015), Winner-take-all which uses recurrent inhibitory connections (Hylander et al. 1993), Hopfield network (Weinfeld 1989) and associative memory based models (Graf and De Vegvar 1987).

2.2.4.3   Stochastic neural network model

These models introduce randomness into the processing of a network either in the activation function or weights which offer a better choice for optimisation problems. The best example of this type is Boltzmann Machine which was used in the early 1990s. But, now a modified version known as restricted boltzmann machine is used, since the training time is significantly reduced and it also forms an integral part of deep belief network (Lu et al. 2007).

2.2.4.4   Unsupervised learning rule model

These models are popular for implementations in neuromorphic systems beyond STDP. They are most popular online learning mechanism and are employed for large datasets. Hebbian learning mechanism (Schneider and Card 1991) and Self organizing maps (Tamukoh and Sekine 2010) fall under this category.

2.2.4.5   Visual system inspired model

These models play a vital role in neuromorphic implementations pertaining to classification and deep learning. Cellular neural networks (Mosa et al.) were the earlier used visual system inspired model whereas, in the early 2000, pulse coupled neural networks (Wang et al. 2010) gained popularity. Other less used network models include Cellular Automata (Isobe and Torikai 2016), Fuzzy neural networks (Kuo et al. 1993) and hierarchical temporal memory based models (Ibrayev et al. 2016).

2.2.4.6   Spiking neural network (SNN) model

The biological neurons in the brain processes binary spike-based information. According to Neuroscience, the timing and number of spikes code the information carried by Neurons (Thorpe et al. 1996). Driven by this observation, the past few years have witnessed fast progress in the modelling and formulation of training schemes for Spiking neural networks as a new computing paradigm that can potentially replace ANNs as the next generation of neural networks (Sengupta et al. 2018).

Spiking neural networks represent a special class of third generation models where neuron models communicate by sequences of spikes. Each neuron maintains an internal membrane potential, which is a function of input spikes, a constant membrane potential leakage coefficient, current membrane potential, and associated synaptic weights. A neuron fires when its membrane potential exceeds the neuron's firing threshold value, which is emitted as a spike to all connected synapses or neurons (Pande et al. 2013).

The Event driven nature and improved energy efficiency are the two major factors that popularised the spiking neural network model. In a spiking neural model, the training is usually done in a traditional neural network and the resulting solutions are adapted to fit the spiking neuromorphic implementations. Hence, there are a variety of SNN models derived based on the traditional network implementations.

The different SNN models include spiking feed-forward network (Ponulak and Kasinski 2011) which is usually applied in low level sensory system, spiking recurrent networks (Diehl et al. 2016), with its rich dynamics and high computational capacity used to study associative memory mapping, spiking deep neural networks (Han et al. 2016), spiking deep belief networks (Stromatias et al. 2015), spiking hebbian systems (Bofill et al. 2002), spiking hopfield or associative memories (Ang et al. 2011), spiking winner-take-all network (Oster et al. 2008), spiking probabilistic network (Hsieh and Tang 2013), spiking random neural networks (Abdelbaki et al. 2000), Central pattern generators (Donati et al. 2014) which generate oscillatory motions and is therefore used in robotic motion.

2.2.5   The Choice of Network Models

The massive interconnection that happens in brain eliminates the concept of understanding the high levels from organisation of low levels. This is because in brain the high level of abstraction evolves from how low level components respond to the interaction with their environment. Hence, while defining the choice of a suitable network model the environment where it is employed also plays a major role. Thus, there is always a multiple choice for the same problem based on the application.

While Stochastic neural network model suits best for optimisation problems, Spiking neural network model is the best choice for classification and deep learning based problems. Recurrent neural network models play a promising role in natural language processing applications which involve more predictions.

2.3   LEARNING ALGORITHMS

Learning algorithms play a vital role in neuromorphic computing which is absent in traditional computing systems. The accurate and faster the learning, the speed of the system arriving at the solution increases.

The best learning algorithm takes into consideration the network topology, model and characteristics of the network. Its choice also depends on whether the training or learning is off chip or on chip and also whether the algorithm is online unsupervised or offline supervised or both.

Most of the learning algorithms are grouped under the two categories namely supervised and unsupervised as in Figure 2.3.

2.3.1 Supervised Learning Algorithms

The most primitive supervised algorithm is back propagation. It can be applied to various network models like feed forward, RNN, SNN and Convolutional Neural Network (CNN).

There are various approaches available for back propagation based learning. They are off line (Haykin and Network 2004), online (Krid et al. 2006), hardware based (Ueda et al. 2014), gradient descent based optimisation (Nair and Dudek 2015) and chip in the loop (Yang et al. 1999).

The two major challenges faced by back propagation algorithms are the restriction on the type of neuron model and network topologies, and the higher cost involved in implementing as hardware. Hence, various other models were developed.

Few of the popular models include LMS Algorithm (Walker and Akers 1988), weight perturbation (Maeda and Tada 2003), learning mechanisms developed for CNN (Fieres et al. 2006), Boltzman machines (Sheri et al. 2015) and hierarchical temporal memory (Ibrayev et al. 2016).

Another category of supervised learning includes evolution inspired algorithms. They don't rely on particular characteristic or a model and optimize within characteristics of a particular hardware.

They can be either on-chip or off-chip based on the principles of Genetic Algorithm (NirmalaDevi et al. 2009), Particle swarm optimization (Bezborah 2012) and differential evolution (Buhry et al. 2009).

2.3.2   Unsupervised Learning Algorithms

The unsupervised learning algorithms are necessary to extract the full potential of neuromorphic systems. They can be implemented on-chip and on-line. Few early implementations were based on self organising rules or self-organising maps (Mann and Gilbert 1989). On the other hand, Hebbian type learning rules (Rachmuth and Poon 2003) are very popular as on-line mechanisms for neuromorphic systems. There are variations in the Hebbian implementations.

However, the most popular one is Spike-Timing Dependant Plasticity (STDP) which is more biologically inspired. STDP also has custom circuits for potentiation and depression (Kanazawa et al. 2003) in synapses. A lot of work needs to be carried out to develop algorithms for neuromorphic systems as learning is one which differentiates this, uniquely from that of Von Neumann based computing. Even algorithms like back propagation and associated memory models were developed with Von Neumann architecture in mind.

2.3.3   Factors that Govern the Algorithm Choice

Hence, the factors to be considered for choice of algorithms for neuromorphic implementations are (i) the chosen hardware, (ii) the chosen model, (iii) online or on-chip or both based learning, (iv) Arrival speed vs accuracy of results and (v) bio-plausible or bio-inspired learning.

With regard to object recognition, supervised learning is still the dominant learning methodology. When datasets grow, having fully labelled dataset is costlier and hence may likely move to semi-supervised and unsupervised methods. This thesis is limited to the use of supervised learning algorithm. In supervised learning, traditionally two kinds of problems suffice

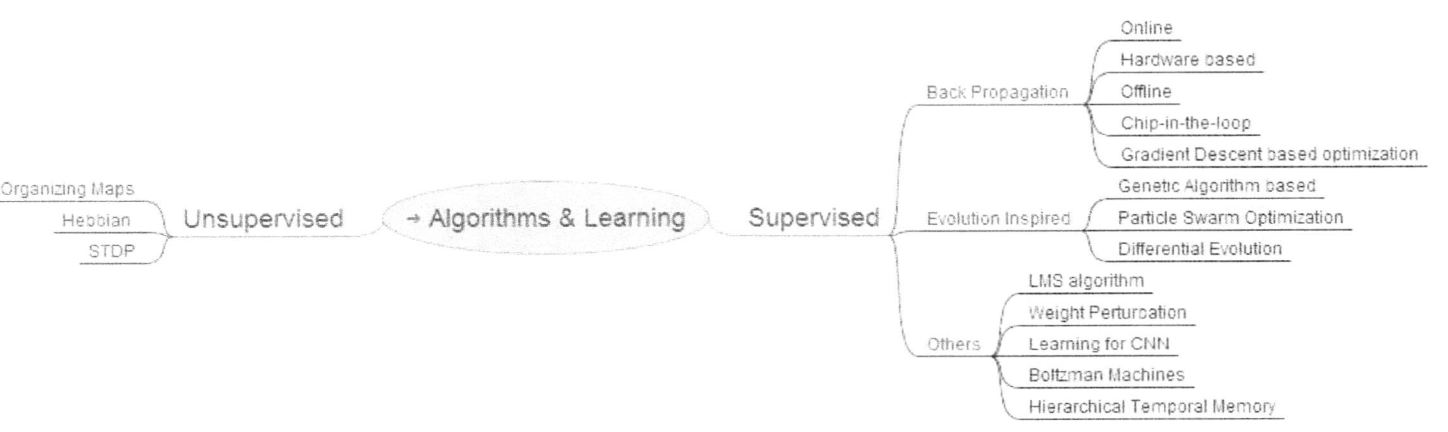

Figure 2.3: A mindmap of Learning Algorithms

namely regression and classification. While regression involves continuous output values like computing a function of input, classification systems compute a single-discrete output. The focus of this research is limited to classification problems which is typically used for object recognition.

CHAPTER 3

DEEP LEARNING WITH NEUROMORPHIC MODELLING - A PROMISING APPROACH TO BIO-PLAUSIBLE SYSTEMS

3.1   INTRODUCTION

Deep learning (DL) is a subset of machine learning algorithms and was based on Artificial Neural Networks (ANN). It consists of multiple layers where each layer transforms its input data into a slightly abstract and more complex representation. Hence it was conceived as limited to abstract applications of ANN (Kay 2017). However kriegeskorte (Kriegeskorte 2015) developed a new framework to apply these methods to the study of the brain.

Neuromorphic modelling, on the other hand, aims at constructing models that are more biologically plausible and is designed to understand the functionality of the brain. Deep learning and neuromorphic systems at the core are concentrated with how a set of neurons carry out the function of some interest by communication through connection weights. While the former is concerned more about information processing, the latter aims to build a suitable architecture for implementing the same. Hence a combination of deep learning and neuromorphic computing provides a promising approach in building more biologically mimic systems and hence took the lead in providing better image classification systems. However, researchers in the respective fields differ from each other. This is because deep learning is generally perceived as nonlinear optimization abstract problems and can't be applied to study the function of the brain (Kay 2017). Hence they are mostly used in ANN and are limited to abstract applications.

As a result, the two fields are studied in isolation. Researchers in DL use tools like Torch (Collobert et al. 2011), Caffe (Jia et al. 2014), Theano (Team et al. 2016), or TensorFlow (Abadi et al. 2016b), as simulation tools for constructing their models. Neuromorphic modellers are more focussed on tools like NEURON (Hines and Carnevale 1997), NEST (Gewaltig and Diesmann 2007), PyNN (Davison et al. 2009), Brian (Stimberg et al. 2013), or Nengo (Bekolay et al. 2014). However, a combination of these tools in practise with the aim of utilising the best of both research fields provides a better solution to classification problem as can be observed in this thesis.

3.2 THE INTEGRATE-AND-FIRE - A BIO-PLAUSIBLE NEURON MODEL

The integrate-and-fire (IF) neuron (Lapicque 1907) was one of the primitive computational models of a neuron. It was developed based on the idea that the neuron membrane can be observed as a capacitor storing charges over time (Abbott 1999b). It does not measure the electrical and chemical changes that happen in a behaving neuron. It was based on two main behaviours:
1) the capacitor integrates current over time and
2) the voltage on reaching threshold fires.

Additionally, a leaky term may be included as a resistor in parallel with the capacitor that allows charge leakage over time. The integrate-and-fire (IF) is used to refer to the non-leaky model, and leaky integrate-and-fire (LIF) is used to refer to the leaky version.

3.2.1 The Non-leaky Integrate-and-Fire (IF) Model

The functioning of the IF model can be described as follow: The charge Q across a capacitor is given by the equation 3.1 as

$$Q = VC, \tag{3.1}$$

where V is the voltage across the capacitor and C is the capacitance. As current is time derivative over charge, differentiating this with respect to time, the membrane voltage $V(t)$ of the neuron is given by

$$C\frac{dV(t)}{dt} = J(t) \tag{3.2}$$

where $J(t)$ is the input current to the neuron over time, and C is the membrane capacitance.

Hence, an IF neuron shows a simplified model of integrating the input current over time as shown in equation 3.2. Thus the accumulation of charge takes place.

To understand the firing property of neuron, a threshold voltage V_{th} is defined. When the neuron voltage increases this threshold, the neuron fires as described in neurophysiology. The firing process once begun is irreversible in nature. So, it is necessary to define a reset procedure. As a result, once the neuron completes firing a spike, the membrane potential is reset to the resting potential V_{rest}. In neurophysiology, the ionic currents like potassium bring back the membrane voltage to resting potential.

Thus, a basic neuron function is defined using an IF model. Few other IF models include an absolute refractory period in which the voltage is held constant at resting potential for some period after a spike. The length of time in resting potential is described by t_{ref}. This is also a neurophysiological principle as few post-spike ionic currents are strong that it becomes practically impossible to spike for a period of time.

3.2.2   The Leaky Integrate-and-Fire (LIF) Model

The basic IF model is modified since neuron membranes do not act as perfect capacitors rather slowly start leaking current over time. Thus, it tries to pull the membrane potential back to its resting value. Hence, the leaky integrate-and-fire (LIF) model (Lapicque 1907; Knight 1972; Koch 1999) was formulated taking this additional physiological property into account. In LIF, the membrane is modelled as a capacitor and resistor in parallel including the leaky term.

The LIF neuron dynamics equation is thus modified from that of LIF as:

$$C\frac{V}{t} = -\frac{1}{R}(V - V_{rest}) + J(t) \tag{3.3}$$

where R is the membrane resistance, V is the voltage across the capacitor, C is the capacitance, V_{rest} is the resting potential and $J(t)$ is the input current to the neuron over time.

The number of parameters can be reduced by normalising the above equation. By multiplying 3.3 with R, replacing $RC = \tau rc$, then making $\bar{V} = (V - V_{rest})/(V_{th} - V_{rest})$ and $\bar{V}_{rest} = V_{rest}/V_{th}$, the equation becomes

$$\tau rc \frac{\bar{V}}{t} = -\bar{V} + \bar{J}(t) \tag{3.4}$$

where $\bar{V}_{th} = 1$ is the firing threshold of new equation, $\bar{V}_{rest} = 0$, and $\bar{J}(t) = \frac{R}{V_{th} - V_{rest}} J(t)$. Here, it is observed that $\bar{J}(t)$ is a linear transformation of $J(t)$ and the voltage and current in the new equation are unitless quantities which simplifies the mathematics without limiting the generality of LIF model.

3.2.2.1 Determination of spike rate

The timing of the neuron spike for a given current can be described from equation 3.4 whereas most of the times in spiking neuron models, it is the spike rate that is required.

For a constant input current, the analytical firing rate for a LIF model can be easily determined. Let the constant input current be $J(t) = j$. This, assuming no spikes are fired makes equation 3.4 as

$$V(t) = (V(0) - j)e^{-t/\tau rc} + j \tag{3.5}$$

Then, the time taken for the voltage to rise from $V(0) = 0$ to $V(t) = 1$ can be calculated if $j > 1$. Substituting these values in above equation and adding a refractory period and inverting, the spike rate r is given by

$$r = \frac{1}{t_{ref} - \tau rc \log\left(1 - \frac{1}{j}\right)} \tag{3.6}$$

3.3   ENCODING   INFORMATION   IN   BUILDING BIO-PLAUSIBLE MODEL

It is also essential to understand how the brain encodes information to design a bio-plausible neuron based model. Towards this, researchers have proposed a variety of encoding schemes which can be categorised into rate codes, population codes, and timing codes.

3.3.1   Rate vs. Timing Codes

The firing rate in a rate coding is based on the number of spikes a neuron fires over a period of time. Muscle contraction is a classic example of rate coding that the brain involves in. On the other hand, the temporal code is dependant on the time of individual spikes which also plays an important role in areas like early auditory system where the timing of spikes helps localize sound. It is also important in the visual system.

Therefore, the fundamental distinction between rate and temporal codes is that in a rate code, the whole system will function as well or better if all spiking neurons are replaced with ideal rate-based approximations. With temporal codes, the timing of individual spike matters, so they will no longer function properly when using rate-based neuron models.

3.3.2   Population Coding

Both the rate codes and temporal codes are used to describe the individual neurons. When a representation is distributed across many neurons it is difficult to decode from activities of few neurons. Hence to encode taking into consideration a group of neurons, population coding is employed. Population coding is essential since it will increase the degree of redundancy if extrapolation of the idea of rate or temporal coding is carried out.

Population coding, on the contrary, represents each neuron in a different aspect of represented value. As a result, it is easier to distinguish between codes that are synchronous between neurons and those that are not. Hence, a postsynaptic neuron takes coincidence detection of two of its input neurons and fires only if it matches.

Rate coding, on the other hand, takes advantage of the synchrony between neurons in terms of their instantaneous firing rates. Hence, population coding is advantagious when the modelling involves networks like CNN and is used in building more bio-plausible systems.

3.4   EVOLUTION OF DEEP LEARNING METHODOLOGIES

Hubel and Wiesel in the 1960s, while studying the visual cortex of a cat found an interesting phenomenon on its method of pattern identification. Its visual cortex is made up of simple cells and complex cells that fire response to properties like the orientation of edges on the visual sensory inputs (Hubel and Wiesel 1962). Further, they observed that the spatial invariance was exhibited more by complex cells rather than simple cells. This inspired the development of later deep neural network architectures.

In 1968, Ivakhnenko and Lapa were the first to develop DL systems based on networks trained by the group method of data handling (GMDH) (Ivakhnenko 1968). It used Kolmogorov-Gabor polynomials as polynomial activation function and used feedforward multi-layer perceptron type architecture. In 1971, Ivakhnenko described a deep GMDH with 8 layers

(Ivakhnenko 1971). In early papers on DL, the layers are incrementally grown, trained by regression and optimized with the help of a separate validation set using decision regularisation. Apart from deep GMDH networks, the first ANN with DL, which includes the neurophysiological insights is known as the Neocognitron (Fukushima 1980).

The Neocognitron is a purely supervised feedforward, gradient-based deep learning architecture with alternates convolution and downsampling layers. Winner-Take-All based unsupervised learning rules were used and spatial averaging was used in downsampling. Present DL mostly combines convolution neural networks (CNN), max-pooling and backpropagation networks. Convolution concept, replication of weights and sub-sampling was introduced to DL because of Neocognitron.

3.4.1 Convolution Neural Network (CNN) Architecture

Neural networks involving image classification problem faced the challenge while learning, as the network parameters grew with lots of variables due to its computational complexity. Even if CPUs and GPUs got faster, the Von Neumann bottleneck was a major limitation. At this juncture, the architecture of CNN gained momentum by the end of 1990, as the key contribution of the convolutional architecture is the dimensionality reduction. CNN's are a group of neural networks commonly employed for analysing visual imagery.

A CNN Architecture transforms the image volume into an output volume using a list of few distinct types of layers. Each layer accepts an input 3D Volume which is transformed into an output 3D volume by means of a differentiable function. The layers may or may not have parameters and additional hyperparameters. The layers are arranged with the objective of extracting the features of the image.

While using a CNN, feature extraction and classification are the two steps involved. The CNN layers have a set of learnable filters which traverses over the height and width of the input volume to produce a 2D activation map that provides its response at every spatial position. This is done to compute dot products between the entries of the filter and the input at any position. It enables the network to learn any type of visual features such as an edge or a patch of colour, or even patterns on higher layers of the network. In CNN, every neuron is connected only to a local region of the input volume. The spatial extent of such connectivity is called the receptive field of the neuron. A Convolution layer accepts a volume of size W1 x H1 x D1, uses the hyperparameters namely spatial extent F, number of filters K, Stride S and amount of zero padding P to produce an outer volume of W2 x H2 x D2 where W, H and D represent the width, height and depth respectively of the layers. The formula for calculating the spatial size of the output image is given by (W-F+2P)/S+1 where W is the size of the input volume. The three hyperparameters namely stride, depth, and zero-padding control the size of the output volume.

3.4.2   CNN with Spiking Neurons

Convolution neural networks achieved huge success in image classification tasks. For example, a highly accurate handwritten character recognition is achieved with the use of carefully designed layers of neurons as in CNN (LeCun et al. 1990). One important factor for the success achieved in CNN is its implementation mostly in GPUs which are energy rich. However, high energy consumption and computational complexity appear to be the limiting factors for their usage in mobile and robotic applications. Hence, spiking neural networks (SNN), the third-generation neural networks appeared to be an alternate and most promising solution. SNN by its use of event-based spikes in encoding information ensures efficient power consumption and handles better algorithmic complexity.

3.4.3   Spiking Neural Network – The Third Generation Neuron Model

The biological neurons in the brain processes binary spike-based information. According to Neuroscience, the timing and number of spikes code the information carried by Neurons (Thorpe et al. 1996). Based on this observation, various training schemes are formulated and modelled for the SNN over the past decade hoping to replace ANN. Thus SNN was seen as the next generation of neural networks (Sengupta et al. 2018). They use spike driven impulses unlike constant time-variance as output which is the norm in conventional neurons. The information is encoded in the amplitude of the output in the conventional neurons whereas, in case of SNN, it is based on spike rate or spike time. Each neuron has an internal membrane potential maintained as a function of input spikes, a constant membrane potential leakage coefficient, current membrane potential, and associated synaptic weights. A neuron fires whenever its membrane potential increases above the neuron's firing threshold value, which is emitted as a spike to its other connected synapses/neurons (Pande et al. 2013).

SNN, similar to a biological neuron can process temporal patterns along with spatial. SNN exhibits fast real-time signal decoding and allows the possibility of information multiplexing. The addition of temporal dimension enables high information carriage capacity thus scoring high over traditional networks (Paugam-Moisy and Bohte 2012). It has been demonstrated that it is found to exceed the computational power of perceptrons and sigmoidal gates when applied to non-spiking as well as spiking based solution (Maass 1997). SNN has the capacity to emulate the brain in the areas of fault tolerance and can repair itself (Harkin et al. 2009). As a result of this feature, adaptation capability is achieved, which improves network reliability. SNN is able to process an enormous amount of data using relatively fewer number of spikes (VanRullen et al. 2005). SNNs hardware operation is event-driven and thereby process input information only at the moment of arrival of binary spike signal. Under such circumstance, the input spike train is sparsely-distributed which ensures the reduction of power consumption. The features that hold SNN high when implemented in hardware are taking complete advantage of the hardware inherent parallelism and the ability to cater to the demands of the real-time fault-tolerant application.

Bio-plausible nature of spiking models not only offer solutions to applied engineering problems like event detection, classification, robotics control, image processing and as such but also acts as a powerful tool for analysing the elementary processes in the brain which includes learning, neural information processing and plasticity.

3.4.4   Pains and Gains of SNN Implementations

The most important challenge SNN faces is the availability of suitable training data. Since there is a dearth of benchmarked datasets, it involves a process to transform frame-based datasets to frame-free ones and vice versa. The second challenge is the technology readiness level which is at present low as these biologically inspired neurons are more complex and difficult to understand than that of the first two generations. Hence the development of efficient learning algorithms is still at the research level. However, SNN still has high hopes owing to the asynchronous way of processing information which makes it faster and also the event-based activation of neurons which makes it more energy efficient, thus enabling it to be implemented in small processors, unlike its predecessors.

Since SNN processes information in an event-based asynchronous way, evaluation occurs only during activation and, not at every time step, unlike CNN. This reduces the demand for computation. A spiking neuron need not wait to fire its response, for all the neurons in a layer to be evaluated, or till the next discrete time step. This gives the SNN an ability to process information immediately without any delay and is known as pseudo simultaneity (Pérez-Carrasco et al. 2013). Unlike in CNN which works with discrete time steps, the information in SNN is processed as it arrives which makes it faster. Also, since activation is done only on addressing an event, SNN is much more energy efficient than a conventional neuron in a CNN architecture. The availability of supervised and unsupervised learning algorithms where traditional neurons can be substituted with spiking neurons, use of temporal coding and augment weights with delay lines makes SNN implementation more promising. Since SNN can arrive at the results faster and with effectiveness, a replacement of CNN with spiking neurons is chosen as a suitable method for implementation of SN-CNN model. The approach combines the advantage of well-developed learning method of CNN with the improved energy efficiency, asynchronous system behaviour and adaptability of SNN to achieve better and faster results.

3.5   BIOMIMIC   ACHIEVED   THROUGH   NEURAL ENGINEERING FRAMEWORK (NEF)

The NEF is a theoretical framework that leverage single neuron models towards building large-scale neural network models. Based on cognitive abilities, it provides principles aiding the construction of a neural model incorporating its functional objectives, control theory and anatomical constraints. The ability of NEF design is demonstrated by its study in single cell activity (Stewart et al. 2012), behavioural error (Choo and Eliasmith 2010), and response time (Stewart and Eliasmith 2009).

In all these studies, the model is able to match the physiological and psychological objectives without specifically being built into the design. A connection weight matrix that occurs between two neural population computes a non-linear function as proposed by the transformation principle of NEF. This matrix can be divided into two smaller matrices. With the usage of smaller matrices instead of the full connection weight matrix, NEF makes the model computationally efficient that can be made to run on low-cost hardware.

A large-scale neural model in NEF is specified based on three principles of mathematical theory namely representation, transformation and dynamics.

3.5.1 Representation

The information is initially encoded by a population of neurons into a time-varying vector of real numbers. The encoding process is characterised by injecting a specific amount of current into single neuron models based on the vector being encoded. This enables the neuron to spike. During decoding, the original vector can be estimated with enough neurons.

The input signal during encoding drives each neuron based on its tuning curve. The tuning curve is a function of its encoding weight, the gain of the neuron and its bias. As tuning curves are determined for any type of neuron, NEF supports all categories of neurons.

During the decoding process, an economically decaying filter is used to initially filter a train of spikes similar to the way a spike generates a postsynaptic current. The filtered spike trains are summed with decoding weights. The decoding weights are obtained by solving the least-squares minimization problem. Thus, through the use of non-linear function encoding and linear decoding, a population of neurons are collectively represented as a time-varying vector of real numbers.

3.5.2 Transformation

Synapses are connections in neurons that help in communication. They are unidirectional and when a neuron spike occurs, it releases neurotransmitters through the synapse which imparts some amount of current in the postsynaptic neuron. The various factors affecting the amplitude of this current is summarised into a scalar connection weight matrix. This matrix represents the strength of the connection between neurons.

The connection weights between the two neuron population depends on the product of the decoding weights of the initial population, the encoding weights between the postsynaptic neuron population and any linear transform. The NEF hypothetically assumes that the synaptic weight matrix has low rank and is divided into encoders, decoders and any linear transform since in practise the use of full connection weight matrix is rare.

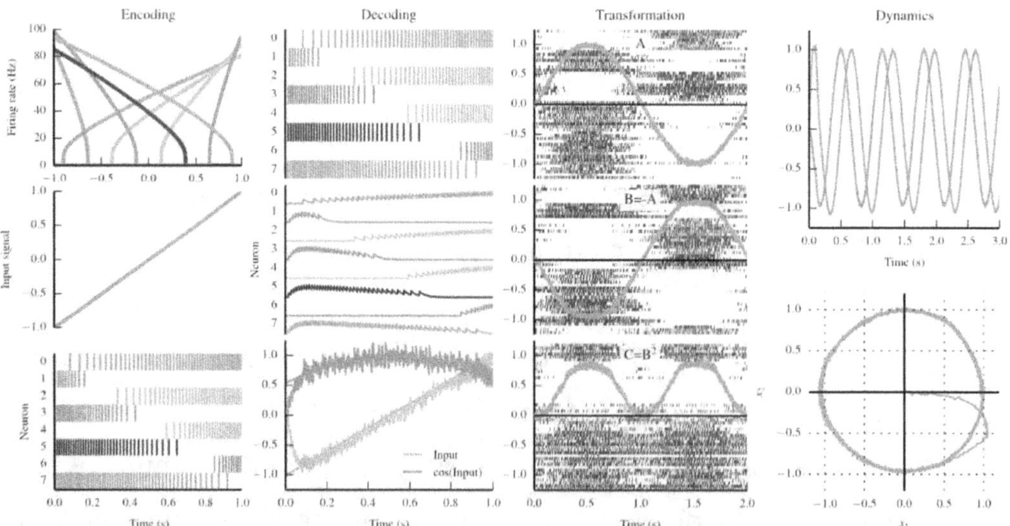

Figure 3.1: Summary of the three principles of the Neural Engineering Framework(NEF) (Bekolay et al. 2014)

3.5.3   Dynamics

For neural systems requiring activities through recurrent connections, the equations that govern the dynamics of that system are analysed using control theory based methods. These are then translated back into the neural system using the principles of representation and transformation. Thus, for recurrent connections, the vectors are described as state variables in a dynamic system.

In this chapter, it is established that a combination of DL and Neuromorphic modelling provids a better approach to developing bio-plausible systems. The section 3.2.1 and 3.2.2 elaborated the dynamics of bio-plausibililty in single neuron structure. The section 3.3 discussed about the information encoding process similar to the brain with the choice of population coding for the neuromorphic system.

The section 3.4 provided an approach of constructing bio-plausible architecture taking into consideration the advantages of CNN and third generation spiking neurons. Finally, the section 3.5 discusses the scaling capability of single neuron models into large scale systems with the use of three NEF principles. Based on these principles, the next chapter explains the construction of a bio-plausible model for the image classification task.

CHAPTER 4

A BIO-PLAUSIBLE SN-CNN MODEL FOR IMAGE CLASSIFICATION

4.1   BASIC COMPONENTS OF MACHINE LEARNING SYSTEM

A typical machine learning system is described in terms of four basic components.

1. Dataset: It is generally a standard set of data available for learning the problem. A supervised learning dataset consists of input and output pairs, whereas unsupervised learning has only inputs. In reinforcement learning, the environment defines the states and those which produce a reward.

2. Architecture: The architecture includes the construction of a learning system, the parameters learnt and the process by which the system is determined.

3. Loss function: It is the term of measure used to evaluate a particular set of parameters. In supervised learning two functions are generally employed, one for training and one for testing. The objective function is also known as loss function which is a measure of performance of the system.

4. Optimising algorithm: The algorithm that minimises the loss is termed as the optimizer. Ideally, the choice of the parameters that

bring the result to the smallest possible loss value on a standard dataset is identified by the algorithm.

Based on the choice of these components, the dataset is fixed, evaluated over an objective function and the components are varied to obtain the best results at the output.

While generalizing to a new dataset, two challenges are faced by ML systems namely overfitting and underfitting. Underfitting refers to the condition when the system is unable to capture the function it is trying to learn. This can happen if the architecture is not powerful enough to represent the function. An example would be a linear network learning on a nonlinear regression function. Overfitting occurs when the model starts learning relationships that are not useful to the problem. This happens if the architecture is too powerful to represent the function. In this case, the learner starts performing on training data but fail miserably on test data.

Hence, overfitting is easier to detect whereas underfitting is not so straight forward. One way to identify is that any network that is not overfitting is underfitting as there is a very minimal window where the network works perfectly. Any network is either always overfitting or underfitting to some extent. Overfitting is not bad but can be reduced for optimal results. Convolution neural network is successful because it makes the parameters easy to learn. Overfitting can be kept on check by reducing the number of parameters with some prior knowledge of the problem. For example, in an object recognition problem, local connectivity in the first layer improves performance as low-level visual features are local.

The values of parameters can be limited by regularization, which prevents learner to be partial on some correlations over others. Drop out is a newer form of regularisation involving probabilistic methods to silence neurons in neural networks. Thus it can be used to reduce overfitting. The other option to help with overfitting is by expanding the dataset.

4.1.1 Objective Functions

The objective or loss function of a network in supervised learning describes the quantity that needs to be optimised relating the actual to the desired outputs of the network. In classification problems, the loss function is different from an error function. An error function quantifies the error made by the network during classification. It typically is the ratio of samples incorrectly classified to the total number of samples. The loss function, on the other hand, might penalize the correct class if the output is not as high as desired. This makes the classifier get correct output by a significant margin thereby making the system more robust and to generalize to new inputs. Another distinction is that the loss function is continuous and differentiable whereas the error function can be discrete. This makes loss function more easily optimizable by the use of derivative.

4.1.2 Dataset

The dataset has two uses: 1) to find optimal parameter values by training the system and 2) to evaluate the values by testing the system. Ultimately, it ensures better performance when exposed to new stimuli. Hence, while using supervised learning the datasets are generally divided into distinct training and testing subset enabling the learner, to be tested with exclusive data it never encountered in learning. The parameters related to architecture, loss function and optimization method are not learned but needs to be chosen. These are known as hyperparameters which include learning rate, size of the model etc.

To support the development of new machine learning models and to compare between models and methods, a variety of standard datasets are publicly available. Many datasets are available in a supervised learning environment for image classification like MNIST, Cifar-10, Imagenet and so on. In this thesis, the standard MNIST dataset and a much more complex FashionMNIST dataset are used to analyse the classification system. The summary of them is shown in table4.1

Table 4.1: Summary of datasets used

Dataset	Image size	# classes	# train	# test	Description
MNIST	28×28	10	60k	10k	Handwritten digits from zip codes
Fashion MNIST	28×28	10	60k	10k	Fashion articles like shirt, pant, etc.

4.1.2.1 MNIST dataset

The Modified National Institute of Standards and Technology database(MNIST) dataset (LeCun et al. 1998) is a collection of handwritten digits (taken from US zip codes written on envelopes). The digits have been preprocessed in a form such that each image contains exactly one centred digit in greyscale (essentially black and white) of high contrast.

The first success in optical classification of handwritten digits by (LeCun et al. 1998) was a turning point in ML systems. Even though the dataset is no longer challenging, it served as a first benchmark for any novel method introduced. Hence, this research is initially tested with MNIST before going for a more complex dataset and the results are verified. Also MNIST allows expedient training even with slow methods which made it quite popular for evaluation of ML classification system.

Figure 4.1: Example images from MNIST dataset

4.1.2.2   FashionMNIST dataset

Fashion-MNIST is a dataset of Zalando's article images. It consists of a training set of 60,000 examples as a set of 28x28 grayscale images, and a test set of 10,000 examples. The training set is divided into 55000 training examples and 5000 validation examples.

The images are associated with a label from 10 classes like T-shirt/top, Trouser, Pullover, Dress, Coat, Sandal, Shirt, Sneaker, Bag and Ankle boot. Each row is a separate image. First column is the label and rest are pixel number with each value depicting the darkness of the pixel. The FashionMNIST shares the same structure of training and testing as MNIST.

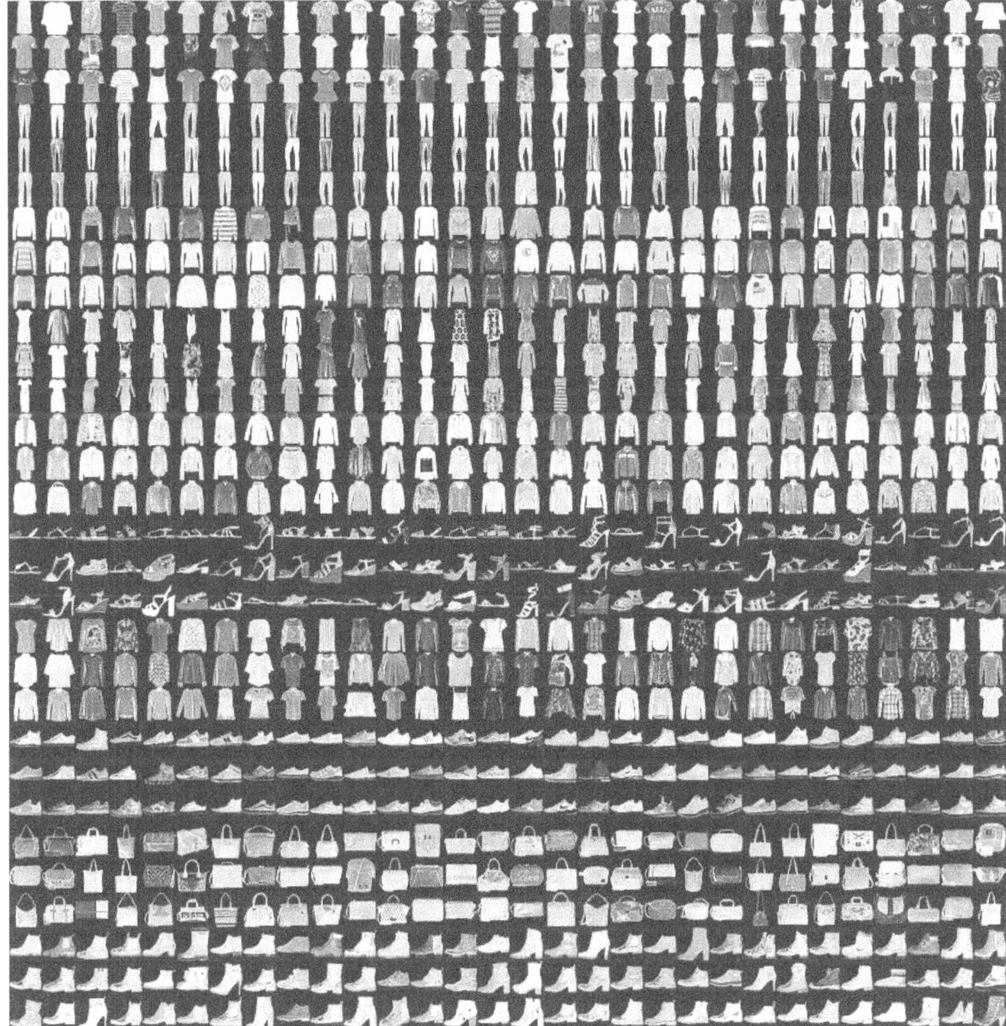

Figure 4.2: Example images from FashionMNIST dataset

4.2   DESIGN OBJECTIVE FOR SN-CNN CONSTRUCTION

A novel spiking neuron model is developed and applied for classification of standard FashionMNIST dataset. The SN-CNN model is developed with the following two objectives. First, the model is to be more biologically plausible. Second, the model while tested in a complex standard dataset must be able to accomplish fairly good performance. The bio-plausibility is achieved by using SNN as it encodes time-varying inputs into trains of perfectly timed spikes. In the same way, the brain works, the input spike trains of SNN when tightly correlated are most likely to fire. The choice of the complex dataset is FashionMNIST as it consists of a more complicated set of fashion articles in ten different classes while retaining the structure to that of standard MNIST.

In order to fulfil both the objectives, the process involves using traditional DL techniques to train static images from a standard dataset. From the training network, the parameters namely the weights and biases are taken and used in the testing network which is formed by replacing with spiking neurons. The test patterns of the dataset are then tested in the spiking neuron network.

The challenge involves altering the training network by varying the input parameters so that the weights and biases are perfectly transferred to the testing network while minimising the classification error between the training and testing network. For analysis, the test dataset is simulated in both the training network involving non-spiking neurons and the testing network with spiking neuron replaced and the results are compared.

4.2.1   Simulation of SNN Model

The SNN model, when simulated with parallel implementation, provides well-balanced communication time and computation cost compared to traditional neural networks as proved by the SpikeNET software (Delorme et al. 1999). During SNN simulation, a spiking neuron does not receive weight values from each presynaptic neuron for every communication step, unlike a rate coded traditional neuron. This helps in removing the bottleneck of message passing as only a few neurons are active at each time in an SNN.

4.3 NENGO - A SIMULATION PACKAGE FOR LARGE-SCALE NEURAL SYSTEMS

Nengo, the software package in which the simulation is carried out is generally employed for simulating large-scale neural systems. The goal of Nengo is to ensure the creation of neural models with the support of NEF principles and connect those components through its object model. It provides an API (and GUI) for modellers to build neural models. The API allows users to take advantage of the NEF while supporting other network methods like deep-learning. Nengo allows these simulations to run on different backend hardware like CPU, GPU, FPGA, and neuromorphic hardware.

Nengo is designed to support temporal dynamics of neurons, build complex neuron models like SNN, complex network structure and provides a capability for the designed model to run across different hardware backend, with a minimal cognitive load on the user(Rasmussen 2018). Nengo was previously used in implementation of models of human motor control (DeWolf et al. 2016), systems of visual attention (Bobier et al. 2014), modelling speech production (Kröger et al. 2014), planning with problem solving (Knight et al. 2016), study of inductive reasoning (Rasmussen and Eliasmith 2014), construction of working memory (Stewart et al. 2010), reinforcement learning (Stewart et al. 2012), as well as large integrative end-to-end functional models of the brain that combine these systems (Eliasmith et al. 2012).

4.3.1 The Architecture of Nengo

In Nengo, ensembles describe the information being represented and the connections describe the transformation happening on that information. These descriptions are implemented as object models and are translated into a network of interconnected neurons. Nengo thus translates a functional model at a higher level to a neural model at a lower level similar to a "neural compiler".

A Nengo model is composed of 5 basic objects namely network, ensemble, node, connection and probe. A network contains other Nengo objects that are functionally related and is described by a group of interconnected neurons. An ensemble represents the neuron group while a node inserts signals into a model and represents non-neural information. A connection is used to describe the connection between nodes or ensembles or both towards information transfer among objects. A probe for analysis extracts data from the model during simulation.

Each of these objects has two major tasks. They initially define the network structure and is used to store parametric information like neuron type, synaptic filter values, connection weights and bias values for the ensemble. Thus a front-end Nengo network is like a large data structure which defines the model parameters and its structure. The back-end then takes the information encoded in that data structure and implements that which matches the specification.

Nengo architecture has a separate front-end model construction and back-end implementation. The front-end code is used to define the structure of the network. The back-end implementation has two objects namely signals which has a generic tensor array of simulation values and, operators which read input, perform some operation and stores the output. Thus the back-end simulates the network on the underlying computational platform.

This division between model construction and simulation enables a Nengo model to be run on multiple simulators (Bekolay et al. 2014). Nengo with its base on the neural engineering framework (NEF) helps in designing the spiking neuron model on which the learning algorithms are implemented.

4.4 NENGO DL: DEEP LEARNING INTEGRATION WITH NENGO

NengoDL integrates the strength of neuromorphic modelling with deep learning through a software framework. It allows users to construct complex and biologically specific neural models and mix them with DL elements like convolution networks. Those models are efficiently simulated in a user-friendly unified framework. In addition, users can apply DL training methods in the biological neural models, to optimize the parameters (Bekolay et al. 2014). A nengo dl.simulator is used to simulate a Nengo network via TensorFlow. The detailed architecture is provided as in Figure 4.3. As NengoDL resides on the backend of the Nengo architecture, the model construction process remains unchanged from the user's perspective while switching from Nengo to NengoDL.

Thus NengoDL resides primarily on the back-end side of the Nengo architecture. In other words, the model construction process is largely unchanged from the user's perspective when switching from Nengo to NengoDL.

In this thesis, a Nengo network is formed using NengoDL and is simulated in a TensorFlow computational framework. While the class NengoDL.simulator is used in simulation and optimizing the parameters, the Tensor node inserts the TensorFlow code into Nengo model. Tensor node allows easy inclusion of models such as convolution network. SNN Toolbox (Rueckauer et al. 2017) is most likely related work to NengoDL functionally. The goal of SNN toolbox is to consider a network constructed in a deep learning package like Theano or Caffe and convert it into a type of SNN that closely matches the performance of the source network. Although this can also be done under NengoDL the advantage is that it supports the simulation as well as optimization of both spiking and non-spiking network in a unified framework,

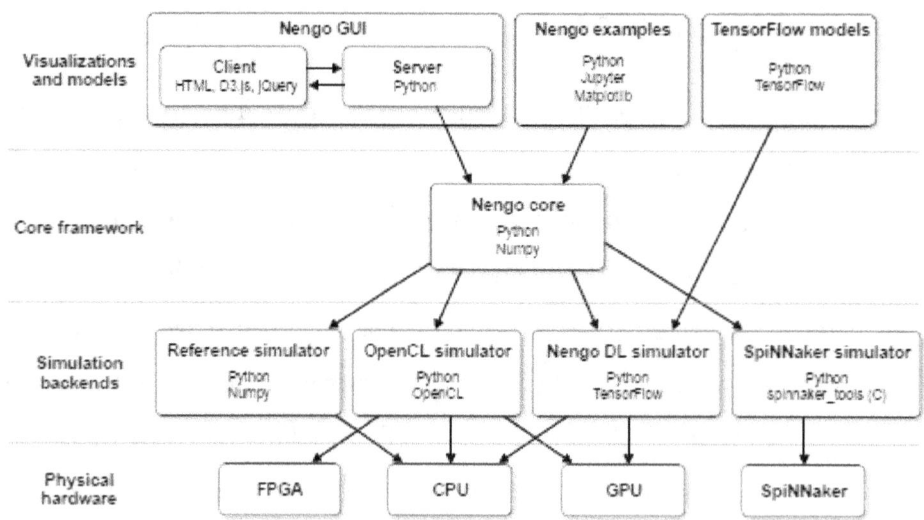

Figure 4.3: Nengo Architecture (Rasmussen 2018)

unlike SNN toolbox. Thus NenogDL provides a more generalized scope than SNN toolbox.

4.5   A DECLARATIVE PROGRAMMING APPROACH USING TENSORFLOW

TensorFlow is an open source machine learning framework developed by Google (Abadi et al. 2016b) and is employed to construct static as well as dynamic networks. It uses data graphs to represent computations, shared state as well as operations that alter the states. It works with a wide variety of heterogeneous systems as the dataflow graph can be mapped across many machines in a cluster, and within a machine across multiple devices like general-purpose GPUs, multi-core CPUs, and Tensor Processing Units (TPUs) that serves as custom-designed ASICs for TensorFlow.

Unlike previous "parameter server" designs where the management of the shared state is built into the system, there is flexibility in TensorFlow for developers to experiment a new and wide variety of algorithms for the deep neural network. It is being used for deploying machine learning systems across a dozen areas of computer science and other fields (Abadi et al. 2016a).

All states in machine learning including mathematical operations, the parameters and their update rules are represented in a single dataflow graph. All data are modelled as tensors (n-dimensional arrays). As dataflow makes communication between sub-computations explicit, it simplifies distributed execution. TensorFlow uses declarative programming approach where the user can specify the computations they tend to perform at a more abstract level.

The user starts by defining some input Tensors and then apply different operations on the input Tensors. The TensorFlow then translates the declarations into actual steps to be executed on the GPU, CPU or any other hardware or simulator. This type of automatic placement determination on a given set of devices in TensorFlow frees users from this concern.

The ability to experiment with a wide range of optimization algorithms, support advanced machine learning algorithms and automatic placement determination are the features that enabled us to choose TensorFlow for simulation. Two key features from TensorFlow ensured that it is employed along with NengoDL in this research. They are

4.5.1   Automatic Differentiation

The declarative graph based programming that assists in automated manipulations of that graph. An element, when added to the graph can be automatically added to all the other elements that require computation on the element. This from a user perspective is easy to apply gradient descent-based optimization methods.

4.5.2   Accelerator Abstraction

The declarative programming style frees the user from looking into the aspects of custom hardware and software optimizations. Once the structure is defined the Tensorflow automatically takes advantage of the available accelerators to run that program. This ensures significant improvements in Nengo models with high simulation speed.

4.6   CONSTRUCTION OF THE NEURAL NETWORK

The construction of SN-CNN architecture for image classification involves two phases. The first phase involves the construction of CNN based training network using the principles of deep learning in NengoDL. TensorNodes which allow the inclusion of convolution connections are used to construct the network. The function nengo dl.tensor_layer mimics the layer-based syntax of DL package in TensorFlow. Tensor_layer builds a sequence of layers, where the output of the previous layer is taken and some transformation is applied to it. It also has the ability to be passed a Nengo neuron type instead of Tensor function. Using this an ensemble implementing non-spiking softLIFRate, a rate version of the LIF neuron is constructed. SoftLIFRate has a tuning curve with continuous first derivative and has a differentiable approximation of LIF. The CNN architecture thus constructed in TensorFlow is based on the suggestions that are researched and suggested by Hunsberger et al. (Hunsberger and Eliasmith 2015).

The SN- CNN consists of the following layers:

1. One or Two general convolution layers: The input to a convolution layer is an input image and the output is obtained as an activation

map. The filters are applied in convolution layer to extract high-level features from raw pixel data, which the model can then use for classification.

2. One max pooling layer: Pooling layers are added to capture the semantic information of the image.

3. Drop out layer: The pooling layer is followed by a drop out layer with regularization rate of 0.4, and a dense layer with 10 neurons, one for each digit target class. Feature extraction is carried out by convolution and pooling layers whereas classification occurs in the dropout and dense layers.

4. Output layer: The output layer in CNN is a fully connected layer, unlike other layers. Here the input from the other layers is flattened and sent so that transformation of the output into a number of classes happens as desired by the network.

5. The output layer uses a softmax regression as loss function, to compute the loss and its gradient is calculated. Backpropagation begins to update the weight and biases for error and loss reduction. A single forward and backward pass constitutes one training cycle.

The layers of the SN-CNN model are as follows

```
Input              32 x 32 x 1
Convolution 1      3 x 3 size, 32 filters, 1 stride
Convolution 2      3 x 3 size, 64 filters, 1 stride
Pooling 1          2 x 2 size, 2 stride
Convolution 3      3 x 3 size, 128 filters, 1 stride
Pooling 2          2 x 2 size, 2 stride
Fully Connected    128 Hidden Neurons
Dropout            Regularization rate 0.4
Fully Connected    10 Output Classes
```

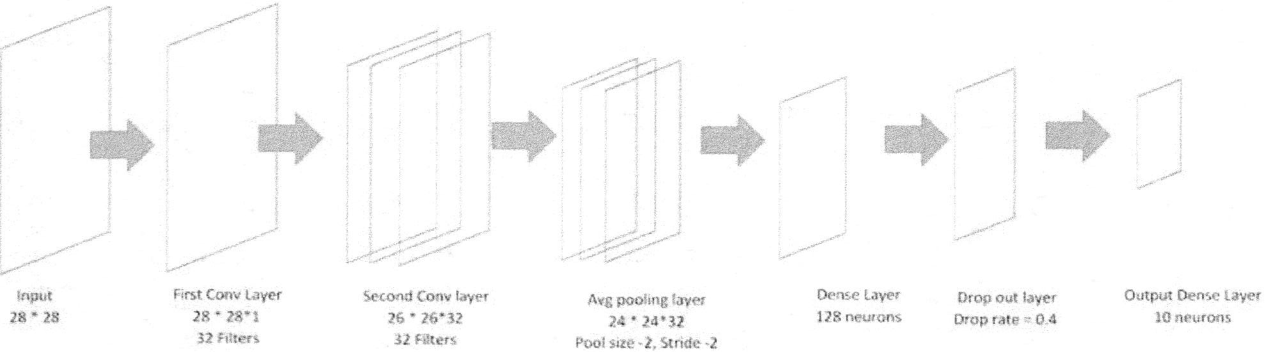

Figure 4.4: CNN Architecture of the SN-CNN Model

4.7   TRAINING AND TESTING THE NETWORK

The input images and target labels from FashionMNIST dataset are initially given to the CNN. A softmax regression is used as a loss function since it is necessary to handle multiple classes.

4.7.1   Choice of Loss Function: Softmax Loss

In softmax regression, the evidence of the input being in certain classes is added by doing a weighted sum of the pixel intensities. The evidence is then converted into probabilities. The weight is positive if the evidence is in favour, else negative. Thus the softmax values for a given image can be used as a relative measurement of the likelihood that the image falls into each target class.

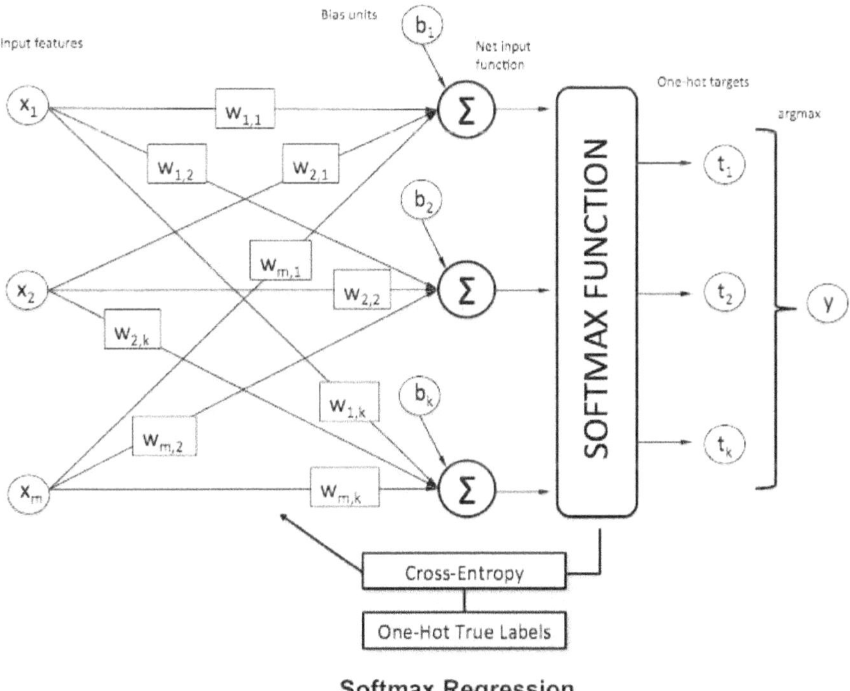

Figure 4.5: Softmax (Win)

The result is that the evidence for a class i given an input x is:

$$\text{evidence}_i = \sum_j W_{i,\ j} x_j + b_i \tag{4.1}$$

where W_i is the weights, b_i is the bias for class i, and j is an index for adding the pixels in input image x.

The evidence is further converted into predicted probability "y" using the "SoftMax" function given by:

$$y = \text{softmax}(\text{evidence}) \tag{4.2}$$

where the SoftMax function for a given class i is given as

$$\text{softmax}(x)_i = \frac{\exp(x_i)}{\sum_j \exp(x_j)} \tag{4.3}$$

Using SoftMax function we exponentiate its inputs and then normalize them.

Softmax loss is the negative log probability of the correct output obtained after passing through a softmax function. A log of the softmax function output has a value normalized by the log of the sum of all exponentiated outputs and is similar to that of linear classifier output. It is more robust across varying number of neurons, unlike weighted square loss. Both softmax loss and hinge loss performs best for large number of neurons, while softmax loss performs better for small neurons also. Softmax loss does not benefit from regularization while moving from rate based neurons to spiking neurons. While considering the time taken to compute in CPU, both softmax and hinge functions scale linearly with the number of neurons. However, Hinge loss takes more time consistently than softmax.

Owing to the benefits of softmax over the weighted loss and hinge loss functions in terms of stability in a robust environment, performance while moving from rate based to spike based and the time taken to compute in low-cost hardware, it is chosen as the error function for the SN-CNN model.

### 4.7.2	Optimising the Weights with ADAM Optimizer

The final step in the Training process is defining the optimizer for training. Adaptive Moment Estimation (ADAM) optimizer is chosen as it combines the advantages of Adaptive Gradient Algorithm (AdaGrad) that works well with sparse gradients and Root Mean Square Propagation (RMSProp) with the typically best performance in an online and non-stationary environment.

Adam is a first-order gradient-based optimization algorithm having stochastic objective functions. It is based on adaptive estimates of the lower-order moment and requires little memory with straightforward implementation methods (Kingma and Ba 2014). Being computationally efficient, ADAM is best for problems involving a huge amount of data and parameters.

The intuitive interpretations of its hyper-parameters ensure little tuning from the user. It calculates an exponential moving average of the gradient and the squared gradient thereby achieving results faster (Ruder 2016). All these factors have led to the choice of ADAM as the best optimizer choice for SN-CNN architecture.

4.7.3   Testing Process

During testing, the network is rebuilt with LIF neurons which makes it dynamic. The saved parameters (weights and biases) from the training network formed using a differentiable approximation of LIF neurons are taken and used in the network formed with spiking LIF neurons. The input parameters are altered and tested to arrive at the best results.

In this chapter, the basic components required for a machine learning system are identified and the dataset choice has been discussed in the section 4.1. standard MNIST and FashionMNIST dataset are chosen for use in the network. The objectives of the bio-plausible model is defined in section 4.2. For choice of simulator, the dynamics behind the selection of Nengo - an spiking neuron based open source simulator, the opportunities to integrate with DL and the declarative learning approach to Tensorflow are presented in section 4.3, 4.4 and 4.5.

The chapter concludes with the description of the construction of an SN-CNN model for image classification as described in section 4.6, the choice of Softmax loss function (section 4.7.1) and the choice of ADAM optimizer (section 4.7.2). The next chapter presents the simulation results of the constructed SN-CNN model in NengoDL and the tuning carried out by various input parameters to improve the results, resulting in the formation of improved SN-CNN model for image classification.

CHAPTER 5

SIMULATION RESULTS AND DISCUSSION

This chapter starts with a discussion of tuning curves of each neuron based on which the function of SN-CNN model can be analysed. The tuning curves provide a visual representation of how well the classification is segregrated for each sample, presenting the input to the network over a time period.

5.1   TUNING CURVES

The tuning curve describes how each neuron responds to an incoming input signal. Here, the input in a single scalar is represented by the x-axis and the response of the neuron is represented in the y-axis. When the parameters of the ensemble are altered, the pattern of the tuning curve is also altered. Thus, we observe from the tuning curve changes, the functioning of the spiking neuron model.

The tuning curve of the LIF neuron and SoftLIF neuron is shown in Figure 5.1. The X-axis of the curve represents the incoming signal input value while the Y-axis represents the firing rate. In the tuning curve, each different coloured line represents the response obtained to one neuron. It can be observed that the neurons cover the space well, whereas there is no clear pattern for their responses. SoftLIF neuron tuning curve has continuous first derivative as it is a rate version of the LIF neuron. The continuity is due to the smoothing around the firing threshold.

It can be generally used as a substitute for LIF neurons in DL during training, and then replaced with LIF neurons when running the network

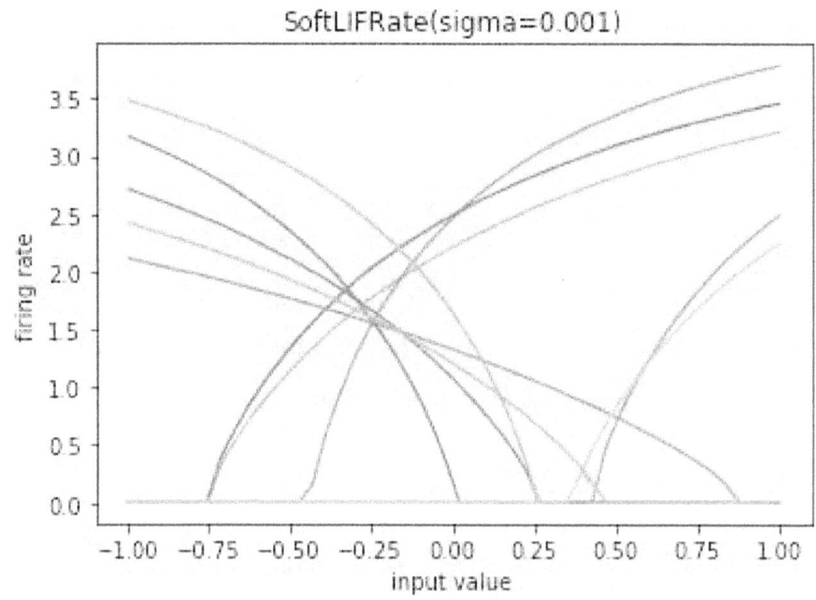

(a) SoftLIF neuron - approximated version of LIF

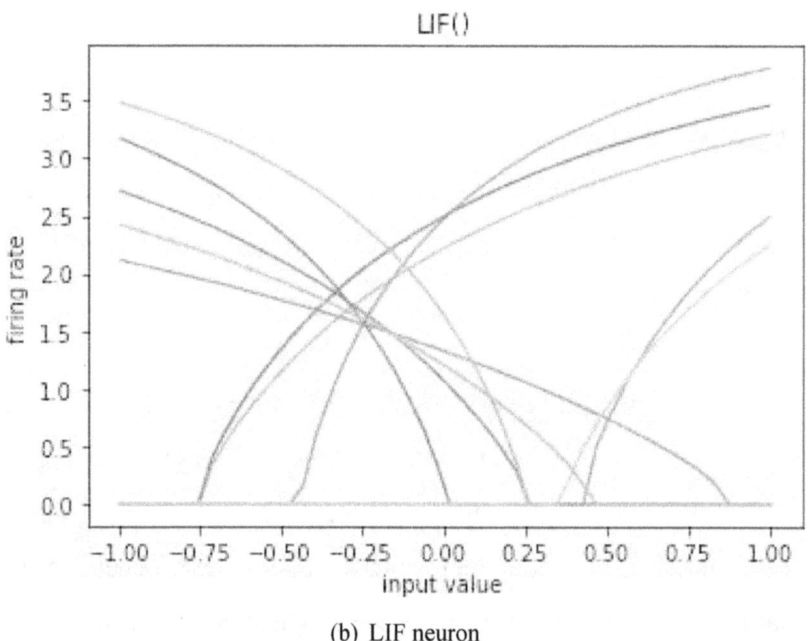

(b) LIF neuron

Figure 5.1: Tuning curve of LIF and SoftLIF for 10 neurons

5.2   PERFORMANCE ANALYSIS OF SN-CNN FOR VARYING EPOCH

The basic SN-CNN model is constructed with better choice of the basic machine learning components mentioned in chapter 4. The model is then tested for different input parameters to arrive at an improved SN-CNN model. The choice of different components are listed below.

CNN Architecture	2 Conv + 1 Pool + 1 Conv + 1 Pool + 1 Dense
Optimizer	Adam
Training neuron	SoftLIF
Testing neuron	LIF
Output Synapse	0.1

The first parameter chosen to be varied and tested is the number of epoch. The results obtained with varying epochs of 10 ,15, and 20 are presented in Table 5.1.

Table 5.1: SN-CNN based classification results for varying epoch

Variable Parameters		Observed Outputs				
Sl. No	No. of Epoch	Error before Training %	Error After Training %	Spiking Neuron Error %	Training Time in hh:mm:ss	Simulation Time in hh:mm:ss
1	10 Epochs	91	6.50	15.00	01:37:26	00:01:48
2	15 Epochs	91.25	9.00	29.50	02:30:23	00:01:48
3	20 Epochs	88.75	7.00	24.50	03:15:26	00:01:49

It can be observed that as the number of epoch changes there is no significant change in the non-spiking or training error. However, as the number of epoch increases, there is an increase in training time.

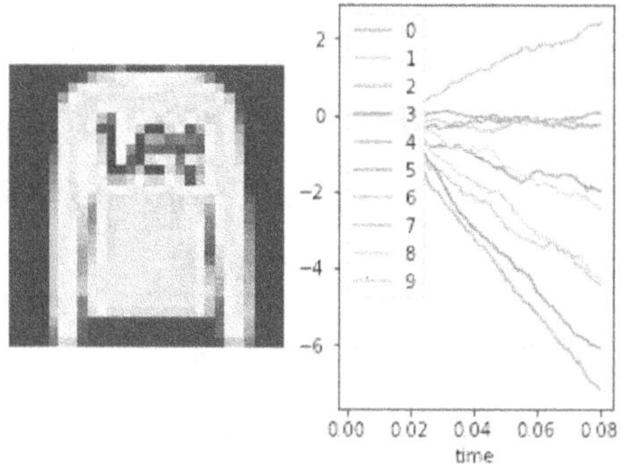

(a) Classification ouput for 10 epoch

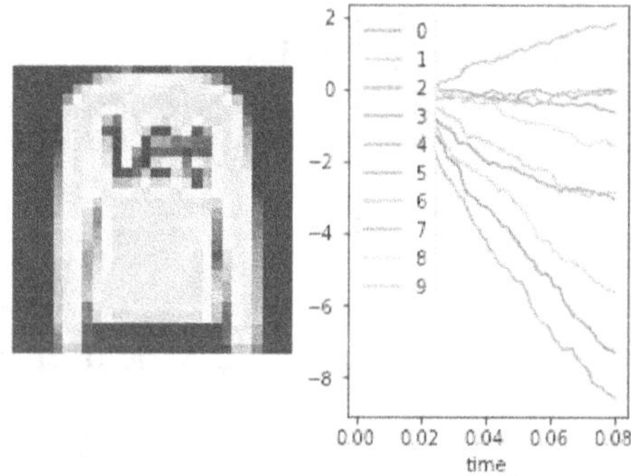

(b) Classification ouput for 15 epoch

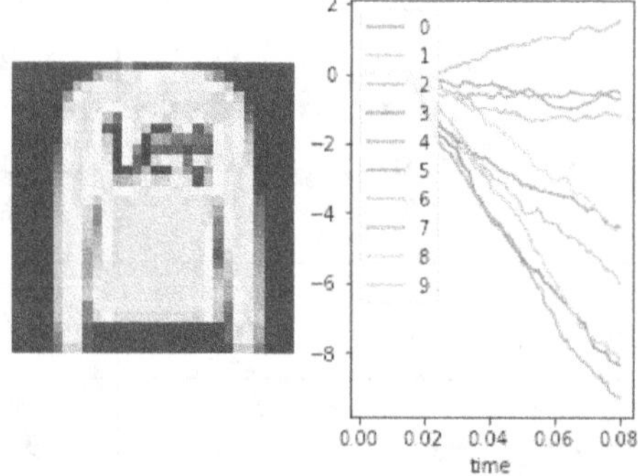

(c) Classification ouput for 20 epoch

Figure 5.2: SN-CNN classification output for single sample with firing graph for various epoch.

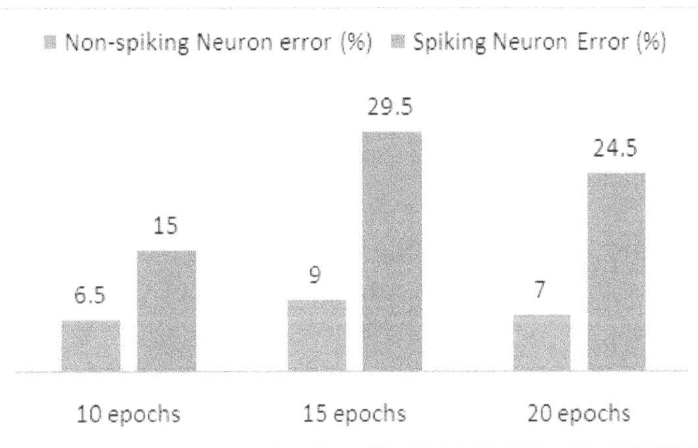

Figure 5.3: Performance of SN-CNN non-spiking and spiking error for various number of epoch

It is also observed that as the number of epoch increases from 10 to 15, there is a double increase from 15% to 29.5% in spiking error even though it subsides further. Hence, taking into account all the observations, 10 epochs is sufficient to test our model and is maintained for further analysis. The other input parameters that are chosen are provided below. The classification output obtained for a single sample with various epoch is shown in Figure 5.2.

5.3   PERFORMANCE ANALYSIS OF SN-CNN FOR DIFFERENT OPTIMIZERS

Two different optimizers are used in basic SN-CNN architecture and the results are presented in figure 5.4. The optimizers chosen are i) generally employed RMSProp optimizer and ii) ADAM optimiser. The theoretical analysis for the choice of ADAM optimiser is explained in section 4.7.2 and is found to provide better results as seen from table 5.2.

CNN Architecture	2 Conv + 1 Pool + 1 Conv + 1 Pool + 1 Dense
No. of Epoch	10
Training neuron	SoftLIF
Testing neuron	LIF
Output Synapse	0.1

Table 5.2: SN-CNN based classification using two different optimizers

Variable Parameters		Observed Outputs				
Sl. No	Optimizer	Error before Training %	Error After Training %	Spiking Neuron Error %	Training Time in hh:mm:ss	Simulation Time in hh:mm:ss
1	Adam	91	**6.50**	**15.00**	01:37:26	00:01:48
2	RMS Prop	91	8.00	19.00	01:37:26	00:01:48

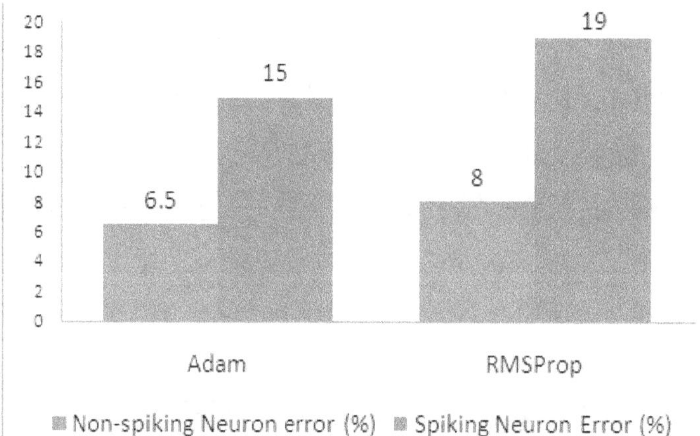

Figure 5.4: Performance of SN-CNN based on non-spiking and spiking error for different optimizers

The ADAM optimizer is providing a spiking error of 15% as against using RMSProp which gives a spiking error of 19% under similar input conditions. Hence ADAM has proved to be the right choice of optimizer for SN-CNN model. The advantage that it combines the benefits of RMSProp and Adagrad ensures it as the best choice.

ADAM optimizer not only provided reduced spiking error but also less error difference between spiking and non-spiking version than RMSProp. All these results, ensured ADAM to be the optimal choice for improved

SN-CNN model. The figure 5.5 gives the firing choice for the optimizers for a single sample.

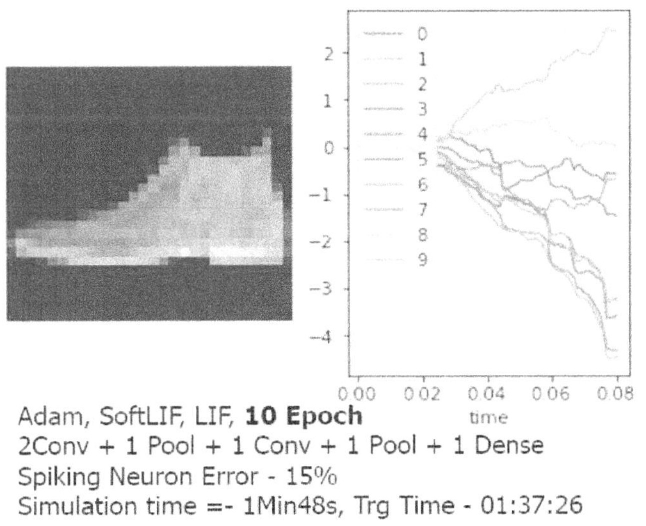

(a) Using ADAM optimizer

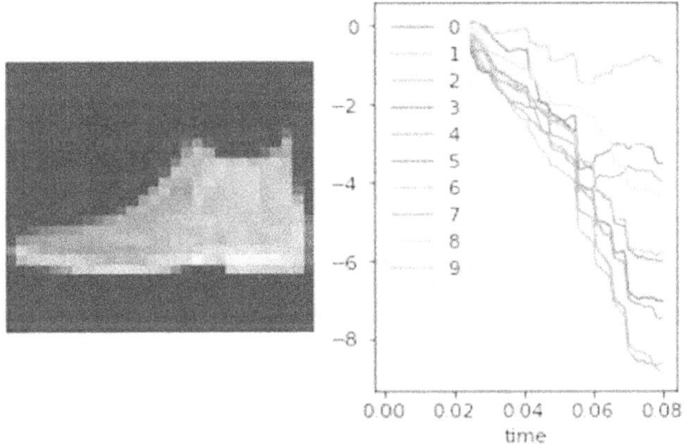

(b) Using RMSProp optimizer

Figure 5.5: SN-CNN classification output for single sample with firing graph for different optimizers.

5.4   PERFORMANCE ANALYSIS OF SN-CNN MODEL FOR DIFFERENT NEURONS

The best choice of neuron was analysed theoretically in section 3.2.2 and LIF was found to be the right choice for SN-CNN model. The SN-CNN model is tested with three different spiking version of neurons namely LIF, AdaptiveLIF and Spiking rectified linear unit (RELU).

It was observed that of these three, LIF has provided better results. Also while evaluating using Adaptive LIF, the rate version of the same is replaced during training and found to yield more spiking error than using SoftLIF. Hence LIF in spiking network and SoftLIF in non-spiking network is found to be the best neuron choice for SN-CNN model.

CNN Architecture	2 Conv + 1 Pool + 1 Conv + 1 Pool + 1 Dense
No. of Epochs	10
Optimizer	ADAM
Testing neuron	LIF
Output Synapse	0.1

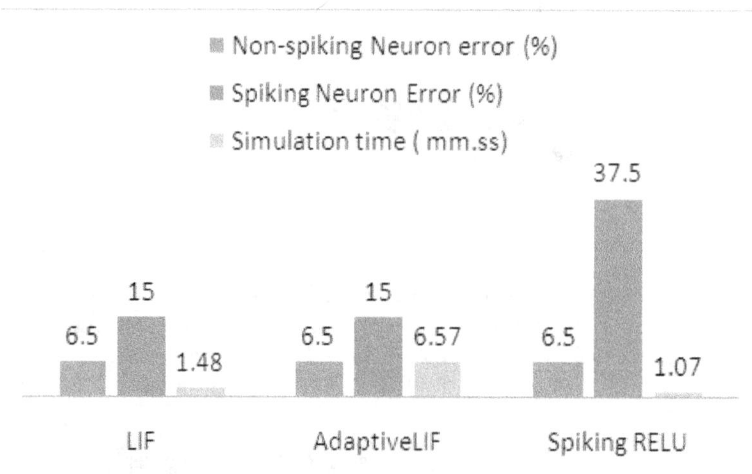

Figure 5.6: Performance of SN-CNN non-spiking and spiking error for different neurons

Table 5.3: SNN based classification using different neuron options

	Variable Parameters		Observed Outputs				
Sl. No	Type of Neuron	Training Neuron	Error before Training %	Error After Training %	Spiking Neuron Error %	Training Time in hh:mm:ss	Simulation Time in hh:mm:ss
1	LIF	SoftLIF	91	**6.50**	**15.00**	01:37:26	**00:01:48**
2	Adaptive LIF	SoftLIF	91	6.50	15.00	01:37:26	00:06:57
3	Spiking RELU	SoftLIF	91	6.50	37.50	01:37:26	00:01:07
4	Adaptive LIF	Adaptive LIF Rate	90.25	12.25	87.00	02:14:55	00:06:28

It can be observed from the figure 5.6 that LIF has reduced spiking error than RELU. In addition to that, it also had reduced error difference when the network was run with spiking and non-spiking versions to a level of 8.5 % which shows that the transfer of weights and biases between them better than RELU. However, when compared with adaptive LIF, LIF neuron outperforms in terms of simulation time as the simulation is carried out under similar conditions. LIF completes simulation in 1 minute and 48 seconds whereas adaptive LIF takes 6 minute and 57 seconds for the same. Hence LIF is proved to be the better choice of neuron for improved SN-CNN model.

5.5 PERFORMANCE ANALYSIS OF SN-CNN FOR DIFFERENT CNN ARCHITECTURAL CHANGES

The basic layers of the CNN architecture are varied without enabling overfitting by forming three different architectures namely basic SN-CNN, improved SN-CNN and Modified SN-CNN.

1. Basic SN-CNN: This model consists of two convolution layers and a pooling layer followed by another convolution layer and a pooling layer. Finally, a dense layer and a drop out layer with a regularization rate of 0.4 is added.

2. Improved SN-CNN: This model has a similar architecture like basic SN-CNN except the intial part has one convolution layer instead of two.

3. Modified SN-CNN: This model is modified version of basic SN-CNN with two convolution layers following the first pooling layer instead of one. Thus it is described as modified SN-CNN with an extra convolutional layer.

All three models are tested with the other parameters similar to them as follows

Number of Epochs   10

Optimizer   Adam

Training neuron   SoftLIF

Testing neuron   LIF

The results are compared and presented in Table 5.4. A graphical comparison of the simulation time and training time for all the three architectures is provided in figure 5.7.

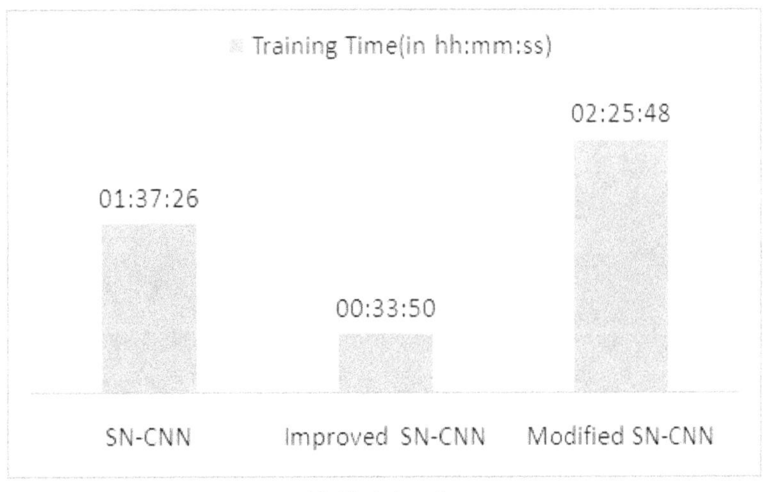

(a) Training time

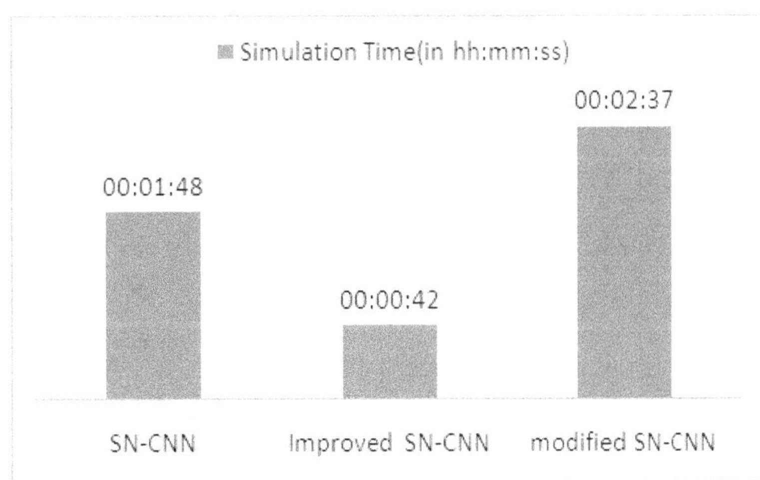

(b) Simulation time

Figure 5.7: Performance of SN-CNN, improved SN-CNN and modified SN-CNN

It can be observed that the best results are achieved with improved CNN model having two alternate convolution and pooling layers followed by a dense layer and a dropout layer with a regularization rate of 0.4.The classification output for a single sample with firing graph involving different architectures is shown in Figure 5.8.

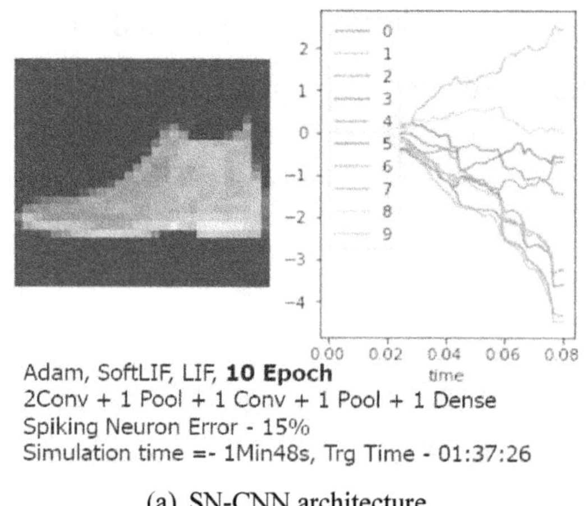

(a) SN-CNN architecture

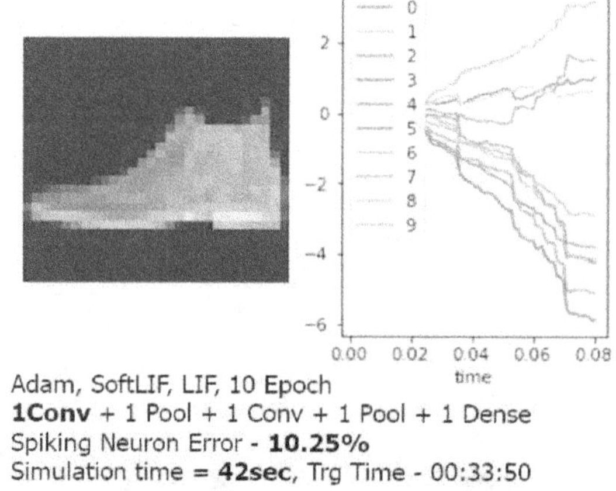

(b) Improved SN-CNN architecture

Figure 5.8: SN-CNN classification output for single sample with firing graph for different SN-CNN architectures.

Table 5.4: SNN based classification for different convolution layers

Variable Parameters			Observed Outputs				
Type of CNN	No. of layers	Output Synapse	Error before Training %	Error After Training %	Spiking Neuron Error %	Training Time in hh:mm:ss	Simulation Time in hh:mm:ss
SN-CNN	2 Conv + 1 Pool + 1 Conv + 1 Pool + 1 Dense	0.1	91	6.50	15.00	01:37:26	00:01:48
Improved SN-CNN	1 Conv + 1 Pool + 1 Conv + 1 Pool + 1 Dense	0.01	89	9.50	10.25	00:33:50	00:00:42
Modified SN-CNN	2 Conv + 1 Pool + 2 Conv + 1 Pool + 1 Dense	0.1	91	8.50	30.75	02:25:48	00:02:37

The error between the training and testing network is shown in figure 5.9. The training network error using SoftLIF was found to be 9.5%. The spiking testing network error after the use of LIF was found to be 10.25%. Since the error difference between the training and testing network is found to be 0.75%, this proves the optimal transfer of weight and biases from the non-spiking to spiking network.

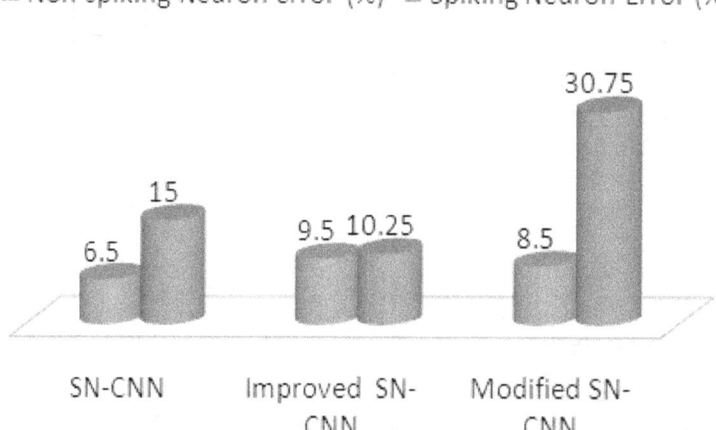

Figure 5.9: Error comparison of SN-CNN, improved SN-CNN and modified SN-CNN

5.6 SIMULATION RESULTS OF IMPROVED SN-CNN MODEL IN MNIST DATASET

The improved SN-CNN model was tested with standard MNIST dataset and the test results for four different samples are provided in figure 5.10 along with their firing graph. It has been observed that after 20 timesteps the output is classified better and has provided a non-spiking error of 0.5% and a spiking error of 1.0%. Hence the improved SN-CNN model is well suited for MNIST also as the difference of error between non-spiking and spiking network is 0.5% which indicated the optimal transfer of weights and biases between them.

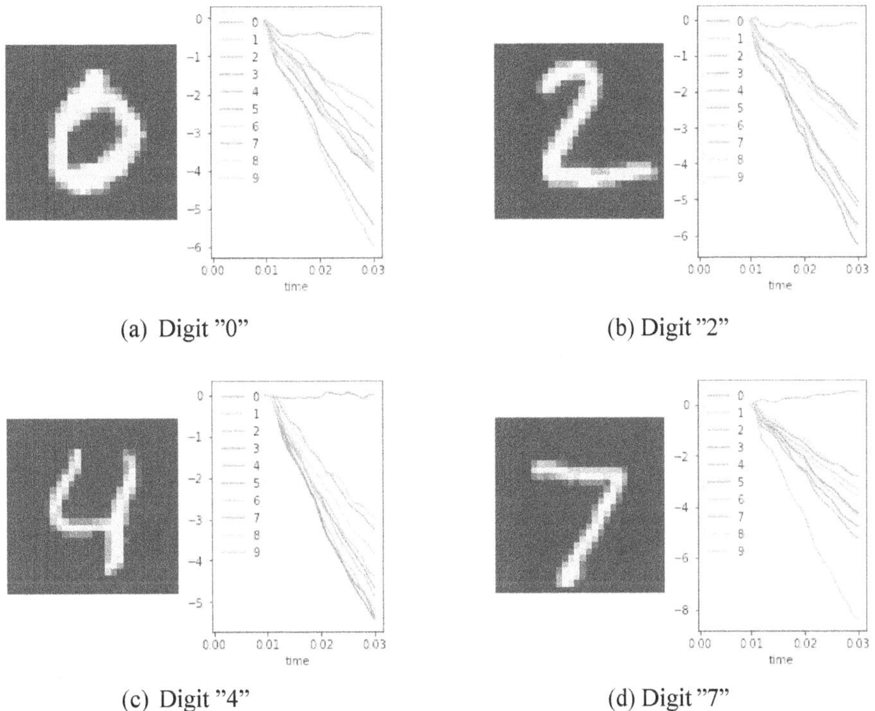

(a) Digit "0"

(b) Digit "2"

(c) Digit "4"

(d) Digit "7"

Figure 5.10: SN-CNN classification output for samples with firing graph for MNIST images.

5.7 SUMMARY OF CLASSIFICATION RESULTS FOR SN-CNN MODEL IN FASHIONMNIST

Table 5.5 shows the various options tried out with different layers, optimizers, testing neuron and a number of epochs. Use of LIF neurons provides best results than Adaptive LIF or Spiking Rectified Linear neurons which are alternates tried in constructing a spiking neuron model.

The model has a convolution layer followed by a pooling layer twice, and then followed by a dense layer provides the best results. In this arrangement we were able to get an error rate of 9.5% during training and with 10.25% Spiking Neuron Error while finishing the simulation of a batch size of 200 in 42 seconds in a normal i5 CPU.

Figure 5.11 provides the results obtained using different variations and their corresponding firing choice.

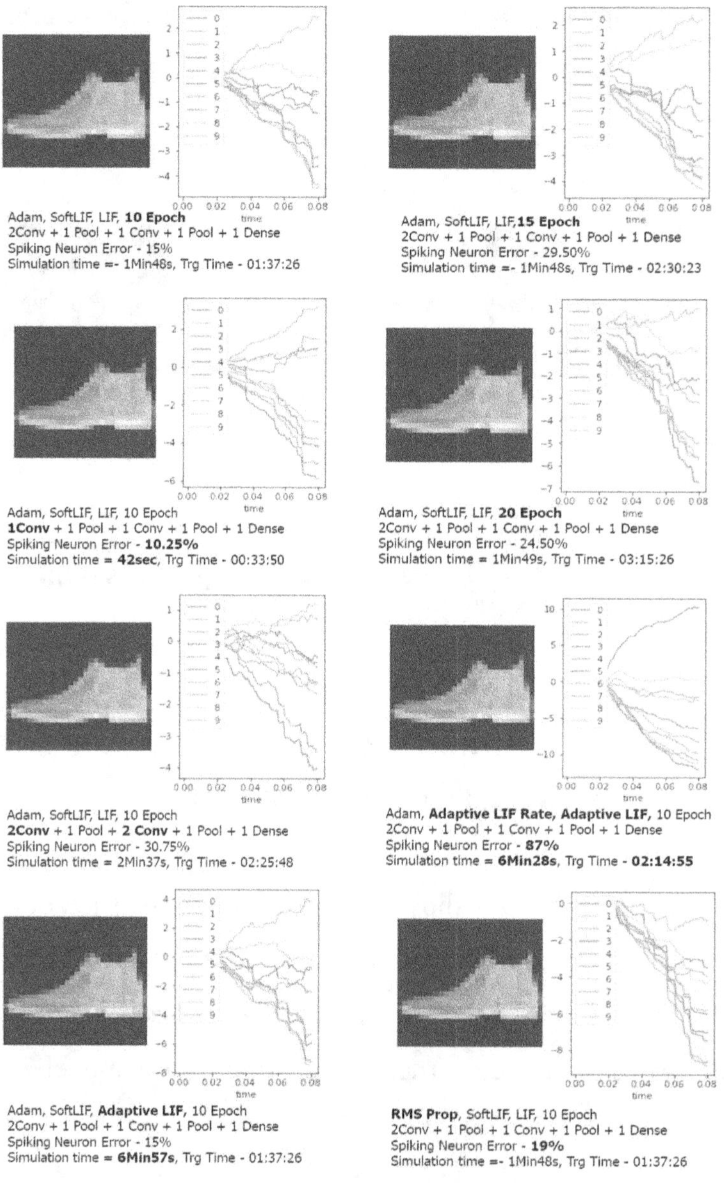

Figure 5.11: Summary of consolidated results towards improved SN-CNN architecture

Table 5.5: SNN based Classification using different input options

	Variable Parameters						Observed Outputs				
Sl. No	Optimi zer	No. of layers	No. of Epoch	Trg. Neuron	Testing Neuron	Output Synapse	Error before Trg. %	Error After Trg. %	Spiking Neuron Error %	Trg. Time in hh:mm:ss	Simulation Time in hh:mm:ss
1	Adam	2 Conv + 1 Pool + 1 Conv + 1 Pool + 1 Dense	10	SoftLIF	LIF	0.1	91	6.50	15.00	01:37:26	00:01:48
2	Adam	2 Conv + 1 Pool + 1 Conv + 1 Pool + 1 Dense	15	SoftLIF	LIF	0.1	91.25	9.00	29.50	02:30:23	00:01:48
3	Adam	2 Conv + 1 Pool + 1 Conv + 1 Pool + 1 Dense	20	SoftLIF	LIF	0.1	88.75	7.00	24.50	03:15:26	00:01:49

Continued on next page

Table 5.5 – *Continued from previous page*

Variable Parameters							Observed Outputs				
Sl. No	Optimizer	No. of layers	No. of Epoch	Trg. Neuron	Testing Neuron	Output Synapse	Error before Trg. %	Error After Trg. %	Spiking Neuron Error %	Trg. Time in hh:mm:ss	Simulation Time in hh:mm:ss
4	Adam	1 Conv + 1 Pool + 1 Conv + 1 Pool + 1 Dense	10	SoftLIF	LIF	0.01	89	9.50	10.25	00:33:50	00:00:42
5	Adam	2 Conv + 1 Pool + 2 Conv + 1 Pool + 1 Dense	10	SoftLIF	LIF	0.1	91	8.50	30.75	02:25:48	00:02:37
6	Adam	2 Conv + 1 Pool + 1 Conv + 1 Pool + 1 Dense	10	SoftLIF	Adaptive LIF	0.1	91	6.50	15.00	01:37:26	00:06:57

Continued on next page

Table 5.5 – *Continued from previous page*

	Variable Parameters						Observed Outputs				
Sl. No	Optimi zer	No. of layers	No. of Epoch	Trg. Neuron	Testing Neuron	Output Synapse	Error before Trg. %	Error After Trg. %	Spiking Neuron Error %	Trg. Time in hh:mm:ss	Simulation Time in hh:mm:ss
7	Adam	2 Conv + 1 Pool + 1 Conv + 1 Pool + 1 Dense	10	SoftLIF	Spiking RELU	0.1	91	6.50	37.50	01:37:26	00:01:07
8	Adam	2 Conv + 1 Pool + 1 Conv + 1 Pool + 1 Dense	10	Adaptive LIF Rate	Adaptive LIF	0.1	90.25	12.25	87.00	02:14:55	00:06:28
9	Adam	2 Conv + 1 Pool + 1 Conv + 1 Pool + 1 Dense	10	SoftLIF	LIF	0.05	91	6.50	14.00	01:37:26	00:01:46

Continued on next page

Table 5.5 – *Continued from previous page*

	Variable Parameters						Observed Outputs				
Sl. No	Optimi zer	No. of layers	No. of Epoch	Trg. Neuron	Testing Neuron	Output Synapse	Error before Trg. %	Error After Trg. %	Spiking Neuron Error %	Trg. Time in hh:mm:ss	Simulation Time in hh:mm:ss
10	Adam	2 Conv + 1 Pool + 1 Conv + 1 Pool + 1 Dense	10	SoftLIF	LIF	0.08	91	6.50	14.75	01:37:26	00:01:48
11	Adam	2 Conv + 1 Pool + 1 Conv + 1 Pool + 1 Dense	10	SoftLIF	LIF	0.01	91	6.50	14.25	01:37:26	00:01:48
12	RMS Prop	2 Conv + 1 Pool + 1 Conv + 1 Pool + 1 Dense	10	SoftLIF	LIF	0.1	91	8.00	19.00	01:37:26	00:01:48

5.8 COMPARISON OF SN-CNN CLASSIFICATION RESULTS WITH CONVENTIONAL CLASSIFIERS FOR FASHIONMNIST DATASET

The usage of SNN helps in reduction of the simulation time while providing almost similar accuracy as that of the non-spiking network. Table 5.6 shows that the improved SN-CNN classification model has been able to provide efficient classification in relation to the various other classifiers applied on FashionMNIST dataset. The ease of implementation in Field Programmable Gate Arrays (FPGA) and other neuromorphic hardware which makes improved SN-CNN model cost effective.

Table 5.6: Evaluation of classification accuracy with other methods

Sl. No	Type of Classifier	Accuracy(in %)
1	Decision Tree classifier*	79.80
2	Extra Tree classifier*	77.50
3	Gaussian NB*	51.10
4	Gradient Boosting classificr*	88.00
5	K Neighbours classifier*	85.40
6	Linear SVC*	83.60
7	Logistic regression*	84.20
8	MLP Classifier*	87.10
9	Passive Aggressive Classifier*	77.60
10	Perceptron*	78.20
11	Random Agressive Classifier*	87.30
12	SGD Classifier*	81.90
13	SVC*	89.70
14	**Improved SN-CNN Model**	**89.75**

(*Results as per Xiao, H., Rasul, K., & Vollgraf, R. et al (Xiao et al. 2017))

From the table, it is observed that the improved SN-CNN model scores good accuracy relative to SVC and gradient boosting classifier. Almost all other classifiers have an accuracy of less than 85% on FashionMNIST dataset. A graphical comparison can be observed in figure 5.12

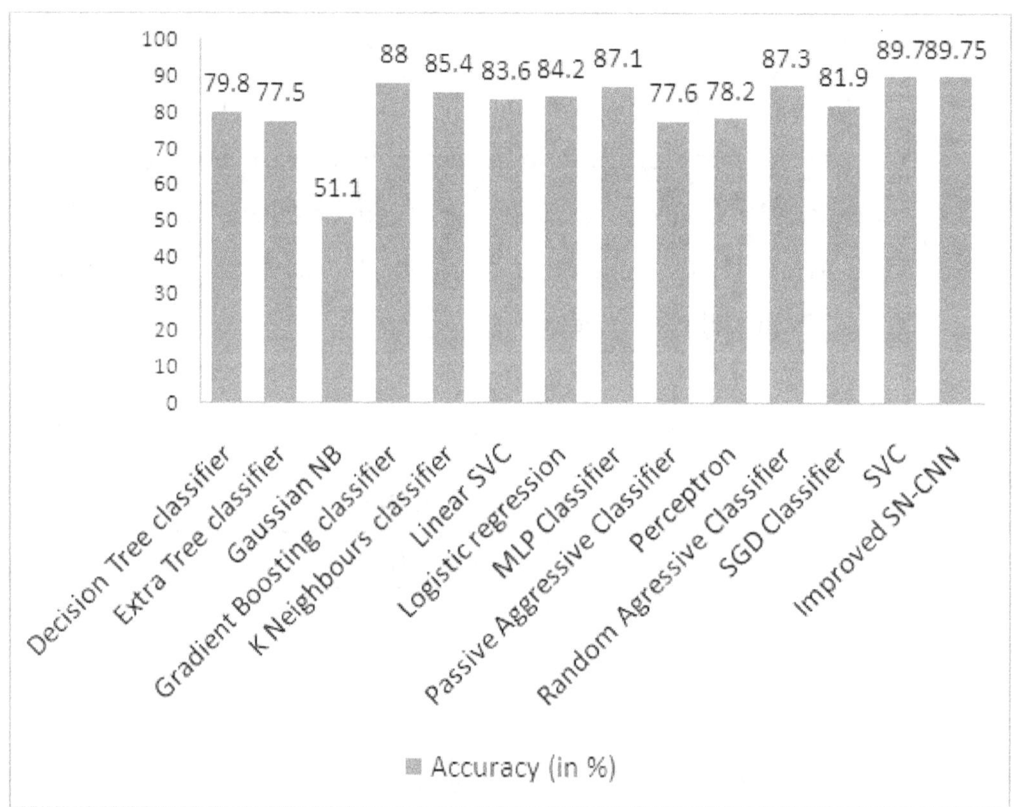

Figure 5.12: Comparison of accuracy of improved SN-CNN architecture with other classifiers

Hence, it is proved that the improved SN-CNN model is able to provide results above the benchmark for FashionMNIST dataset. This was arrived after evaluating the basic SN-CNN in various parameters and construction of the improved SN-CNN with the best results obtained.

The next chapter discusses the opportunities and scope available for implementation of SN-CNN in the best possible choice of hardware.

CHAPTER 6

HARDWARE IMPLEMENTATION ISSUES FOR SPIKING NEURAL NETWORK IMPLEMENTATION

6.1   INTRODUCTION

With the change in the use of computing devices from desktops to wearables, there arose a necessity for innovative implementation of traditional computing (Leong and Jabri 1992). The existing Von Neumann Architecture based sequential computing has been further pushed to the far end of spectrum owing to the saturation of Moore's law. Moore's law has lost its sheen, as it focussed more on technological advancements like increasing integration density by device size reduction, circuit cleverness, etc. Nowadays, the functionality of algorithms has taken priority in a trade-off between clock speed and device feature size. This bought back into focus the research on Neural Networks to a considerable extent in the recent decade.

Hence more focus is provided on mimicking the way brain works, as the application of recent years have moved from being computer based to communication-based. This led to the drastic advancement in Neural science, Neural network-based computing which focuses on parallel processing similar to the pattern of the brain. Brain-inspired computing also known as Neuromorphic Computing focuses more on low power operation (Leong and Jabri 1992), inherent parallelism (Salam 1989), real-time performance control (Hasler and Akers 1990), fault tolerance (Akers et al. 1989), scalability and online learning (Schuman et al. 2017).

However, all the existing popular hardware systems were built on Von Neumann Architecture which provides a major challenge in the implementation of bio-inspired Systems. Hence specialized hardware architectures were developed with a focus on neuromorphic computing. Few surveys on neural network hardware have been published in the recent past (Heemskerk 1995), (Liao 2001).

Neural networks according to their computation can be classified into three different generations. The first generation uses Mc-Culloch Pitts neurons also referred to as perceptrons or computation units. Multi-Layer Perceptrons (MLP), Boltzmann Machines and Hopfield Nets come under these categories. They are characterized by digital input and output. A Multilayer perceptron with a hidden layer computes every boolean function. They generate an output which is "1" or "0". On the other hand, the second generation neural networks are computed based on an activation function which generates continuous output values to a weighted sum of inputs (Maass 1997). Recurrent Sigmodal neural nets and feedforward networks are some examples of second-generation Neural networks. These are universal for analog as well as digital computations and support learning algorithms based on gradient descent. Compared to first generation neurons they are biologically more realistic as the output of the sigmoidal unit is viewed as the current firing rate of a biological neuron.

6.2   HARDWARE IMPLEMENTATION ISSUES WITH NEURAL NETWORKS

The Hardware for second-generation Artificial Neural Network are classified by System architecture, inter-processor communication network, degree of parallelism, numerical representation and typical neural network partition per processor (Schoenauer et al. 1998). NN hardware referred to as "Hardware Neural Networks (HNN)" face a lot of challenges when implemented. To mention a few, (i) It faces stiff competition from general purpose hardware which are already commercially viable, (ii) A high time-to-market as the system design requires good knowledge about algorithms and (iii) Non-availability of good programming interfaces, (iv) Mapping issues on computation on regular two dimensional surface and (v) Introduction of computational error, loss in learning and accuracy due to hardware constraints. These errors indirectly increase the number of cycles required towards convergence on the learning process. On the other side, the ANN-based real-time problem solving, requires capabilities greatly exceeding that of workstations. Simulation of large networks employing more complex neuron models is extremely difficult in general purpose hardware. Hence proper mapping of the network onto the hardware platform is essential for efficient use of parallel hardware.

6.2.1 Parallelism

To understand the issues with hardware implementation, it is necessary to look into the structure of a neural network. Neural Network research is more than five decades old and hence a lot of different families and structures have been developed. But the basic features include a set of neurons in a layered form using one or two hidden layers. The neurons in a layer generate different outputs with the weighted sum of inputs based on a non-linear activation function. Thus the operation of a neuron can be interpreted as a vector-vector product over a function where one of the vectors is input and other is weighted vector. This ensures that the whole layer is considered as a MIMO system where both the input and output are vectors. Hence the relationship between input and output of a NN can be described by a matrix-vector product.

A constant matrix-vector multiplication can be implemented on hardware exploring its inherent parallelism. It is this property which enables high computational speed and fault tolerance for a NN in hardware. The degree of parallelism in digital hardware is related to the number of processing elements involved. Hence parallelism increases the chip area or chip count which makes it more expensive. As a trade-off highly parallel systems usually employ simple processing elements. More efforts are taken to map neural networks efficiently on parallel computers to achieve maximum performance. This is achieved by load balancing, minimizing synchronization between PEs and minimizing inter-PE communication. Further, the mapping should be scalable for both different numbers of processing elements and different network sizes. Based on these, neuron parallelism, synapse parallelism or pattern parallelism is used.

6.2.2 Modularity

The neurophysiology of the cerebral cortex has been known for many decades as having a modular structure. The current models developed and observation studies for the cerebral cortex indicate the need for rich modular units. ANN, especially large scale involves a greater role of modularity in their design. The complex networks are designed using simple and fully connected building blocks. The different categories include decision-based NN, model-based NN, adaptive learning architecture and a structural mixture of experts (Caelli et al. 1999). In certain cases, the processing stages like filtering, feature extraction, segmentation and image analysis are performed by different blocks. In models with parallel structure, each block commits for a particular task. The final results are then consolidated from these individual processing blocks. This type of modularity is known as functional modularity.

In contrast, the structural modularity works with assigning an implicit function to each module. The network structures are divided into a hierarchical set of subnetworks based on the learning and search methods used. Few examples of networks that use structural modularity include Kohonen's self-organising map, nearest neighbour, self-partitioning network and hierarchy based associative memory map.

6.2.3   Dynamic Adaptation

The key feature of the brain is that it's neural coding dynamically adapts to the environment changes as observed in the neurobiological studies. Hence for better neural information processing adaptation emerges as an important feature. Adaptation is the basis for learning in ANN as the weights are adapted as and when it learns. The network parameters are altered according to the pre-defined learning rules. These rules are derived from the error measures or inspired from biological systems making it dynamic. Thus ANN systems without human intervention, have the inherent capacity to dynamically change their architecture and learning rules.

6.2.4   Data Representation

Precision in ANN is to be handled carefully as the reduction of precision decreases the performance of implementation while unnecessary precision wastes the resources. To address this, the right choice of computer arithmetic is essential as it plays a key role in implementation on hardware.   Full parallel implementation of ANN is highly difficult due to the technical limitations of the realizations of bit-level parallelism which increases connections. Also implementing parallel multipliers increase the number of connections which on implementation on chip consumes the larger area. To address these implementation issues, data representation involving bit serial computations and Canonical Signed Digit (CSD) encoding provided promising developments in hardware implementations (Szabó et al. 2000). A study and optimized implementation of square root, multiplication, exponent, logarithm and non-linear activation function in hardware were carried out by Skrbek (Skrbek 1999), who proposed an architecture based on shift-add neural arithmetic involving only adders and barrel shifters.

6.3 ANN IMPLEMENTATION IN DIGITAL HARDWARE

To address these issues various architectures and technologies were studied in the past which include digital, analog, hybrid, and optical. Among them, the digital implementation stands out owing to its advantage of offering high accuracy, low noise sensitivity, greater flexibility, better testability, compatibility with different processors and better repeatability. Mapping the requirements of ANN in hardware with the features of existing hardware architectures is essential to have the right choice of hardware.

Data representation, degree of parallelism and modularity which forms the basic need of Artificial Neural network implementation can be better achieved, if a proper trade-off in terms of performance, flexibility and power dissipation are achieved in the hardware. Figure 6.1 represents the various hardware technologies available for ANN implementation and their levels on performance, flexibility and power dissipation.

The digital HNN implementations are classified as DSP-based, ASIC based and FPGA based. While DSP involves sequential operation and disturbs the parallel architecture of the neurons in a layer, ASIC implementations have a constraint on user reconfigurability. Parallelism within a layer and dynamic reconfiguration are a necessity in most training algorithms. Hence Field Programmable Gate Array (FPGA) is suitable hardware, as it preserves the parallel architecture of the neurons while maintaining the flexibility in reconfiguration.

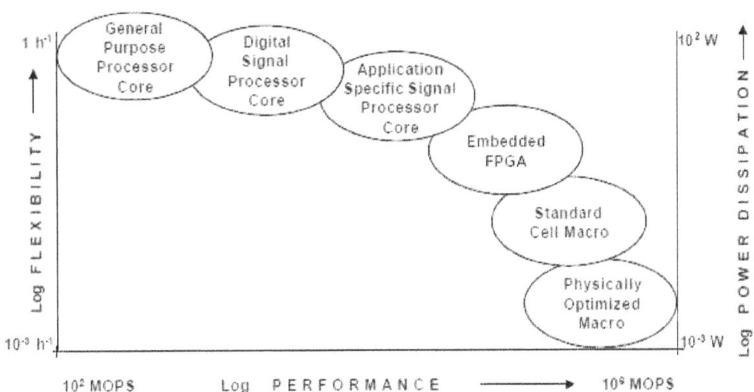

Figure 6.1: Performance, flexibility and power dissipation with various hardware

6.3.1   FPGA Structure - a Recap

FPGA is a combination of many Configurable Logic Blocks (CLBs), configurable input/output blocks, configurable routing blocks and interconnect wires in an integrated circuit.   Each CLB can create a small combinational or sequential digital circuit by configuring the available few look- up-tables (LUTs), adders, multiplexers, and flip-flops. LUTs are mainly used to design shift register or small RAM blocks. These reconfigurable blocks can be programmed electronically by a user to create any arbitrary digital circuit after manufacturing. Recent devices have more complex built-in cells like distributed static RAM and FIFO blocks, multiplier blocks, IP cores, DSP cores, Hi-speed IO blocks, and a higher number of configurable logic blocks (CLBs). New devices have a capacity of up to two million logic cells.

6.3.2   ANN in FPGA

Any ANN is associated with mostly three computational characteristics namely parallelism, modularity, and dynamic adaptation. FPGA are promising in supporting these characteristics through their concurrent processing and rapid reconfiguration ability. Zhu and Sutton (Zhu and Sutton 2003) studied the decade long progress in FPGA implementation of Neural networks. They observed that FPGAs hardware flexibility allows rapid prototyping for initial simulation by time-multiplexing an FPGA according to the sequential steps of ANN algorithm or ANN circuits, topology adaptation owing to its dynamic reconfigurability. Further FPGA provides a variety of data representations through bit-stream arithmetic. However, the implementation issues include a trade-off on weight precision with implementation cost especially in the weight x input multiplication, expensive implementation of non-linear sigmoid transfer functions, requirements of hardware for the conversion of analog inputs and control inputs to digital.

Muthuramalingam et al (Muthuramalingam et al. 2008) presented the implementation aspects of single neurons in FPGA and a comparison of the serial versus parallel implementation in the number of lookup tables used, bit precision involved and computation blocks used. A generalized formula to calculate the speed and resource in slices is also determined.

Skrbek (Skrbek 1999) presents a fast neural computation architecture involving shift-add neural arithmetic. B R Gaines (Gaines 1969) has studied the stochastic ANN model implementation in FPGA. It was also observed that the demand for multiplication in large ANNs are significantly reduced by bit-serial stochastic computing.

Szabo et al. (Szabó et al. 2000) have provided a novel bit-serial distributed arithmetic for pre-trained networks and implemented digital filters based on them. Their matrix-vector multiplication using canonic signed digit(CSD) and bit-level pattern coincidence in the optimization algorithm scores better.

6.3.3   Challenges with ANN Implementation in FPGA

The following are the two major challenges faced by ANN implementation in FPGA.

6.3.3.1   Challenges in ANN structure

Although great empirical studies and observations in ANN are available for FPGA implementation, they are found not as successful as their biological counterparts (Bishop et al. 1995). This is because complex electrochemical signals cannot be modelled using simple equation as in ANN. Further, the neglect of temporal dynamics of the signals and processes in the artificial neural network models (Zomaya 2006) contributed to this situation as in the case of The McCulloch-Pitts model (McCulloch and Pitts 1943) and sigmoid threshold (Bishop et al. 1995) neuron models.

6.3.3.2   Challenges in mapping FPGA architecture to ANN

Modern reconfigurable circuits come with greater gate density with more than 2 million gates. However, efficiently utilising the resources for parallel computing approaches for ANN has not progressed much. Mapping complex, irregular and highly interconnected neural network topologies on regular two dimensional and simple hardware interconnect fabrics of FPGA still proves to be a challenge for ANN.

Reconfigurable computing in FPGA discusses these factors and the tradeoffs between them. Most studies in the field of reconfigurable computing mentioning its benefits is devoted to issues related to harnessing better power dissipation and minimising power issues. They try to improve parallelism and fault-tolerance in reconfigurable platforms with a focus on architectural modifications. They lack in taking into consideration the effects of bio-inspired approaches.

For example, towards obtaining a scalable solution to power wall issue it was suggested to handle with a better trade-off between the number of using low-power processors in parallel and energy (Freeh et al. 2005). However, the brain uses billions of neurons and synapses which are extremely low power and slower in processing. Hence, by employing effective bio-inspired techniques towards self-adaptation and fault tolerance, cheap, highly-dense and larger digital systems can be designed.

## 6.4	BIO-INSPIRED SYSTEMS

No single algorithm is able to universally perform better on all problems. Traditional approaches to computing also face similar situations while solving problems on optimization, pattern recognition learning owing to the use of fixed, precise and deterministic algorithms. Bio-inspired computing on the contrary due to its adaptation principle performs better on a wide range of problems in changing environments. Adaptability in these systems exhibits itself over various time scales and different aspects. Almost all natural systems can be observed by the different forms of adaptation. They tackle more difficult problems by evolving through generations and hence are scalable. They reorganize their resources and recover from resource failure owing to the fault-tolerance available. This open-ended adaptation to changes in ecosystem removes the necessity of human analysis and maintenance and makes them self-sufficient (Shayani 2013). Thus self-sufficiency, automated design and autonomy of system adaptation are the important features that electronic systems need to address to become more biologically plausible.

6.4.1   Spiking Neural Network Implementation in Hardware

A Broad Spectrum of technologies are available for SNN hardware namely CMOS with and without a combination of memristive or special devices, Digital Signal Processors (DSP), Field Programmable Analog Arrays (FPAA), Graphics Processing Units (GPU), Field Programmable Gate Arrays (FPGA), Tensor Processing Units (TPU), accelerators and others.

However, investigating custom hardware solution becomes necessary as during implementing large-scale SNN networks in conventional hardware, large simulation times becomes a constraint.

SNN implementation in digital hardware has various classifications. They are based on

a) Communication method as direct mapping, memory-based and network-based.

b) Requirement of the type of storage as centralised, distributed and massively distributed.

c) Programmability as parametric, software-based, hard-wired or reconfigurable.

d) Arithmetic involved as parallel or serial.

e) Representation of arithmetic as fixed-point, floating-point or integer.

f) Type of architecture used as general purpose, supercomputer, GPU based, customized embedded processors, ASIC accelerators, DSP based accelerators or FPGA based accelerators.

With the variety of these choices, the resulting design can provide different levels of flexibility, scalability, fault-tolerance, speed and capital cost or energy efficiency. However, the right choice of architecture affects many other choices.

6.4.2   FPGA and SNN - the Perfect Match

FPGA is designed as the advancement of VLSI fabrication technology with a view to allowing rapid reconfiguration. Towards achieving this they are made to work in an asynchronous way allowing parallel and distributed processing. These features allowed them to be considered best suited for bio-inspired computing. With a single FPGA, it is possible to make a network of small specific purpose processors and achieve fine-grained parallelism. As a result, they achieve high computational throughput than multi-core general purpose processors.

The memory bottlenecks that appear in GPU architectures are avoided by the use of distributed architectures of FPGA. The advantages of FPGA involving parallelism, short time-to-market and lower nonrecurring costs outshine the disadvantages in clock speeds and power consumption. Dynamic reconfiguration to modify a portion of the circuit while others are working suits them best to be employed for real-time environments. This property promises its usage in a spiking neural network where learning is dynamic in nature. Also, fault tolerance required for SNN can happen in a distributed and parallel architecture as in FPGA.

FPGA happens to be the best choice of implementation in case of Spiking Neural Networks owing to the relationship it has in implementing bio-plausible systems. The table 6.1 depicts the characteristics of FPGA and their relationship with bio-plausibility.

A variety of FPGAs are available that best match the SNN features. Based on application, the right choice of FPGA which makes it more appropriate for implementation can be easily identified. FPGA in terms of computation drastically differs from traditional Von Neumann computing or even GPUs or

Table 6.1: Matching FPGA characteristics with Bio-plausible requirements

FPGA Features	Bio-plausible requirements
Hardware cost	Inversely propotional to compactness
Scalability	Depends on number of neurons, synapses etc
Performance	Simulation and evolution speed
Reliability	Robustness and Fault-tolerance
Design time, Testing time and Complexity	Inversely related to simplicity of algorithm

DSPs that are specially designed for high-end computation. This opens new possibilities for computation optimisation as the SNN hardware architecture differs for reconfigurable devices.

FPGA is highly optimal while involving simple processors with massive parallelism than large computation blocks with many multiply-add arithmetic and distributed RAM blocks. Mapping of SNNs to FPGA has attracted many researchers owing to its reconfiguration ability and interconnects (Ghani et al. 2006) (Glackin et al. 2005). The availability of efficient low-power synaptic junctions and neuron interconnect ensured that scalable SNN models were developed. Since FPGAs have the ability to cater to the adaptable requirements of Neural networks, they enable the fast acceleration of SNNs and hardware-in-loop training. The availability of FPGAs with increased logic and embedded processors popularised this approach.

According to Johnston et. al. (Johnston et al. 2005), SNN when compared with RBF (Radial Basis Functions) and MLP (Multi-Layer

Perceptron) neural networks was most efficient in terms of usage and handling of hardware resources on FPGA.

Thus FPGA emerges as the most prerferred choice when it comes to implementation of spiking network based SN-CNN in hardware when low cost and low volume is a key requirement. However, Neuromorphic systems which are designed from scratch using bio-plausibility can also emerge the best choice towards implementation when it comes to specific applications.

6.5 EVOLUTION OF NEUROMORPHIC SYSTEMS

Brain inspired computing is as old as computer science itself. Replication of biological neural systems in computers has led to key discoveries in the field of artificial intelligence, machine learning and artificial neural networks. The concept of Neuromorphic systems was conceived and published by Turing in 1968 through his work of artificial neurons, the first known modifier device. Later, Chua in 1971 postulated "memristor" the idea to develop a memory resistive device.

However, it gained more importance in 21st century only after HP labs developed memristor as a device and explored its use as synapse in neuromorphic systems. Two main differences hindered the development of neuromorphic systems. First, in a nervous system the processor and memory are the same whereas in Von neumann systems it is not. Second, nature's computing systems are volatile, their states decay, heal and build continously unlike conventional computers. But when the scalable future of Moore's law came to a saturation level owing to Von neumann bottleneck, researchers started trying out alternate options with bio-inspiration in mind.

Non-silicon approaches to material science faced theoretical limits and is only slightly better than CMOS. Neural computing on the other hand, has promising solution since, it is based on algorithmic and architectural change unlike Von Neumann. There has been tremendous interest shown in neuromorphic computing especially in the last decade as we can find more than 1000 research papers reported since the year 2015, compared to an equal amount in the past three decades before (Schuman et al. 2017).

The key motivators for neuromorphic systems are low power operation (Leong and Jabri 1992) inherent parallelism (Salam 1989), inherent fault tolerance (Akers et al. 1989), ability to mimic neuroscience by creating devices with small footprint, real time performance control (Hasler and Akers 1990), speed of computation (Burr 1991), scalability and online learning (Schuman et al. 2017).

The evolution of customizable hardware like FPGAs accelerated the work on evolving a variety of brain inspired network topologies, algorithms and models for neuromorphic computing. The collective contribution by researchers from different fields like neuroscience who studied the computational sense of brain, electrical and electronics engineers who transformed the studies to electrical models, the material scientists who were instrumental in fabricating and characterizing new materials, and computer engineers who developed network models, algorithms and support software, helped in the development of neuromorphic computing to a greater extent, particularly in the last decade.

6.5.1   Neuromorphic Hardware Implementations

Hardware taxonomies for implementation of neuromorphic systems at a higher level can be categorized into analog, digital or mixed signal implementations. The broad categories of digital systems include FPGAs which are most frequently used owing to its ease of programmability

(Berzish et al. 2016). It has more than 300 research papers published. Another category of digital systems namely Application Specific Integrated Circuit (ASIC) designs, involve customized hardware chips like TrueNorth (Hsu 2014), SpiNNaker (Khan et al. 2008), having approximately 200 research publications.

Analog Systems also has a similar division focusing on programmability ease known as Field Programmable Analog Arrays and other on custom analog implementations. A part of FPAA's are specially developed for Neuromorphic systems like Field Programmable Neural Array (FPNA) (Farquhar et al. 2006) and NeuroFPAA (Liu et al. 2008) having inbuilt programmable components. Mixed analog/digital systems owing to its natural similarity to biological systems are also very common. They are used to implement processing components of synapses and neurons. Because of strong reliability to noise environment usually the storage is made up of digital components in mixed systems.

Programmability and learning also are carried out digitally and the rest is made of analog components. Since analog circuitry that operates in sub-threshold mode is highly power efficient, they play a greater role in mixed system design for the neuromorphic system. Two most popular neuromorphic implementations of mixed systems are Neurogrid (Benjamin et al. 2014) and BrainScaleS. While Neurogrid has a closer resemblance to the biological behaviour of neuromorphic systems, BrainScaleS operates at a much higher rate than Neurogrid (Furber 2016).

While GPU's and DSP's act as good accelarators, their choice to use in Neuromorphic systems are limited by their architectural restrictions. FPAA's are still at research level but are promising players in development of Neuromorphic hardwares.

To select appropriate hardware choice the following set of selection criteria needs to be analyzed. They are cost, popularity, type of interface, performance in terms of speed, size and scalability, power consumption, dynamic reconfiguration, fast reconfiguration, data communication bandwidth, validity checking, observability, reliability, and ease of use (Harkin et al. 2009).

6.6 BEST CHOICE OF HARDWARE TECHNOLOGY FOR SN-CNN IMPLEMENTATION

Considering the nature of implementation of SN-CNN model, it is highly necessary to go for an implementation which involves more modularity, high parallelism and ease of implementation. As far as simulation is concerned, the SN-CNN model has shown better results when implemented in CPU. This ensures that the SN-CNN model can achieve better results in low commodity hardware.

Since distributed processing is also possible in SN-CNN model FPGAs can be the best choice when it comes to low volume implementations. However, if the application needs high volume of devices then custom designed neuromorphic systems can serve better.

As described in the above study, since FPGA and SNN match together better than other hardware technologies available, our preferred choice for implementation of SN-CNN is FPGA. However, the integration between the simulator NengoDL to FPGA is still at the research level and it has been found from the developers that it is under development.

Hence, the possiblilities of implementation is only analysed while the actual implementation in FPGA will be possible when the simulator hardware integration is available.

6.7   DEVICES FOR NEUROMORPHIC COMPUTING

Carver Mead first noted that CMOS transistors operating in current mode below threshold of transistor gate voltage have highly similar sigmoidal current voltage relationships as do neuronal ion channels and consume less power. This led to the development of neuromorphic silicon Neurons (SiN) which is analogous to a neuronal function. It offers an intermediate between biological neurons and digital computers with respect to power and space efficiencies. It is also faster than real neurons in terms of computing speed. Further, the computational delay is independent of network size. Already a number of dedicated processes for manufacturing low-power sub threshold SoC circuits are available.

However SiN faces the following challenges (i) high susceptibility to mismatch in transistor threshold voltage (ii) variability in intrinsic VLSI process that limits the scalability, (iii) limitation in input voltage dynamic range in traditional sub threshold current-mode differential pair that compounds the mismatch errors as size increases, (iv) these models are non-mechanistic making adjustment of parameters post fabrication or during computation difficult. Addressing these challenges led to the advent of Tri-Gate process which adds new dimension to VLSI device scaling.

Other approaches use nanotechnologies such as nanowires, memristors etc. Still they are facing problems like non-robustness and less efficiency in terms of yield. Hence, neuromorphic modelling should shift the focus to developing devices taking into account their degree of biological realism and robustness, along with integration density and computation and/or power efficiency (Poon and Zhou 2011).

From Turing's artificial neurons and Widrow and Hoff ADALINE in the 60's, specific devices for neuromorphic computing has evolved to great levels. The key factors that helped in the development of circuit level components are memory technology and optical components. The various device level advancements for neuromorphic computing can be listed as follows

6.7.1   Memristors

The most primitive device specially for neuromorphic computing was called memristor postulated by Chua in 1971. In 2008, HP Laboratories produced the memristor (Yang et al. 2008) also known as "Memory Resistors" and explored their use as synapses in neuromorphic circuits (Snider 2008).

The memristors mimic the synaptic connections of a human brain as their resistance depends on its past states. Since 2008, there has been tremendous development related to device models and use of memristors as alternative computing architectures. Memristors became popular as they exhibit STDP like behaviour and is energy efficient (Strukov et al. 2008).

6.7.2   Conductive-bridging RAM (CBRAM)

These are non-volatile memory technology and uses electrochemical properties to form and dissolve connections. They are used to implement synapses (Burr et al. 2017) and neurons (Jang et al. 2016). CBRAM devices exhibit gradual switching characteristics which is essential for updating weights during unsupervised learning when implemented in hardware.

Yuhan et al. (Shi et al. 2018) has demonstrated high accuracy on character recognition by using CBRAM in unsupervised learning environment. The use of CBRAM provide systems capability to learn and process information autonomously in power restricted environment.

6.7.3   Phase Change Memory

A PCM is a high performance non-volatile memory where phase changes are carried out by applying or drawing an electric current. They can perform unlimited number of writes unlike NAND based conventional non-volatile memory. These can achieve high density than other devices. Hence, they are used to implement synaptic weight storage (Burr et al. 2015), synapses and neurons (Tuma et al. 2016) in Neuromorphic systems.

6.7.4   Floating gate Transistors

These are used (i) as analog memory cells for synaptic weights or parameter storage (Fujita and Amemiya 1993), (ii) to implement mechanisms like STDP (Brink et al. 2013), (iii) to implement dendrite model (Brink et al. 2008) and (iv) to design a neuron (Aunet and Hartmann 2003).

6.7.5   Optical Components

These are used because of inherent parallelism, speed of arriving at results and is less complex in programming. However the major challenge of using optical components is storage (Collins and Penz 1989).

6.7.6   Synaptic Transistors

The Synaptic transistors are specifically designed to simulate the neuron behavior unlike other technologies which focus on memory. The source and drain are used to read the synaptic weight whereas the gate uses conductance to deliver it to a channel. An electrolyte solution modulates the channel's conductance providing the analog behavior central to neuromorphic system.

This technology is still at an early research stage while it aims to accelerate computing speed while reducing power consumption.

Figure 6.2: Devices and technology of neuromorphic systems

6.8   DEVICE CHOICE FOR SN-CNN MODEL

Memristors are the most popular devices used for neuromorphic systems owing to its development by various industries and researchers. They are energy efficient and exhibit plasticity and learning mechanisms similar to human systems.

Strukov et al. (Strukov et al. 2010) has provided a review of the works on hybrid CMOS/memristor circuits. In that, it was found to combine the reliability, flexibility and good functionality present in CMOS systems with high density thin film nanoscale switching devices. It was also demonstrated to have better performance for applications beyond the scaling limits of conventional VLSI systems thus openly increase the possibility for neuromorphic system implementations.

Jason and Bingjun (Cong and Xiao 2011) have introduced a memristor-based reconfiguration FPGA architecture (mrFPGA). The programmable interconnects use only memristors and metal wires to enable interconnects fabricated over logic blocks thereby significantly reducing the overall area and delays in interconnect. Buffer insertion in interconnects achieved more speed with inserted buffers optimized on demand unlike fixed buffer pattern in conventional FPGA. The results tested showed mrFPGA achieving 2.3x more speed, 5.5x area efficiency and 1.57x power savings.

Thus, owing to the availability of the memristor based devices in FPGA and custom neuromorphic hardware implementations, and their performance improvement, memristors have still been the best choice of device for implementing neuromorphic systems. Based on the above discussions, memristor based FPGA system is suggested to be the best choice for implementation of SN-CNN classification model.

CHAPTER 7

CONCLUSION

In this thesis, a novel classification model involving a bio-plausible approach was designed, simulated and the results obtained were analysed. While the ultimate goal of any bio-plausible method is to develop better object classification based on the insights of how the brain solves the problem, each has its own benefits and limitations. This chapter outlines the key contributions of this thesis towards the design of an efficient bio-plausible image classification problem.

7.1 SUMMARY OF CONTRIBUTIONS

The summary of contributions of this research is presented in three areas namely biological learning in image classification system, development of spiking deep network and analyzing the possibility of implementing them in the best suitable hardware. The future scope on these areas are presented in the next section.

7.1.1 Biological Learning - Opportunities and Scope

The first portion of the thesis explains the study of computational neuroscience and network models in the context of machine learning. A review of the literature showed that bio-plausible approach to classification task has been researched over the last decade to a large extent. The availability of a variety of neuron models, synapse models, neural network models were consolidated as a proof of the options available for the researchers to choose, as the best choice of their interest. Hence, it provided a lot of scope for further development of models using biological learning methods.

7.1.2   Spiking Deep Networks

In chapter 3, the possibility of developing a neuromorphic system using third generation spiking neuron was analysed and for a better classification based on bio-plausibility, CNN was considered as the best choice. Further, the requirement of the choice of the neuron was carried out and LIF neuron was found to be the right neuron.

While looking into the possibilities of the loss function, it is observed that the softmax function was better and the optimizer that suited better for the development of SN-CNN is identified as ADAM. The best framework for the development of the classification model was derived from NEF.

With the analysis of various components needed for image classification architecture towards achieving bio-plausibility, chapter 4 focuses on the development of the SN-CNN architecture using spiking deep networks. The design objectives were defined and expanding the single compartment model to a large scale neural network was analysed based on NEF principles. Towards achieving this, NengoDL an open source spiking neural simulator is chosen and the model is constructed using the simulator. The use of TensorFlow extended the implementation of the SN-CNN model to a distributed environment. The SN-CNN model thus developed is made hardware independent owing to the automatic placement of Tensorflow while not compromising on the speed of the simulation.

### 7.1.3	Hardware Implementation of Bio-plausible Systems - An Analysis

In chapter 5, the limitations involved in the implementation of ANN in different hardware platforms are presented and the need for neuromorphic hardware is also analysed. Based on the challenges, the opportunities and availability of hardware were studied and presented for implementation of SN-CNN model. It was found that the model can be integrated better with FPGA and other neuromorphic systems. Also, the scope for implementation in low commodity hardware systems is found better owing to the difference observed in design than other conventional classifiers. Also since the possibility of parallel and distributed processing is heavily embedded in the architecture of SN-CNN model, hardware implementation probability in low-cost devices is found to be very high.

### 7.1.4	Performance Analysis of SN-CNN Model

Chapter 6 elaborates the performance analysis of the SN-CNN model. It was observed that the improved SN-CNN model provided error in terms of implementation in standard MNIST and FashionMNIST dataset as 1% and 10.25% respectively. Also under different circumstances and for various input parameters, the model was tested and the empirical results found to match with the analysis in the previous chapters. The results obtained were compared with the other conventional architectures of FashionMNIST dataset and the improved SN-CNN provided better or similar results in terms of error while exceeding the limitations in implementation constraints in hardware and speed of arrival of results. Thus, the SN-CNN model emerged as a better bio-plausible, spiking neural network based approach, with the advantage of a universal hardware implementation for solving an image classification problem.

7.2   FUTURE WORK

7.2.1   Biological Learning- Opportunities and Scope

The analysis presented was limited to building network models and learning methods. However, there are a variety of concepts of computational neuroscience that is beyond the scope of this thesis. If they are included and analysed, then the learning possibilities can be drastically improved. Also, it helps to narrow down the scope to increased bio-plausibility. This opens up the possibility of combining more applications and developing better universal systems.

7.2.2   Spiking Deep Networks

Spiking deep networks is yet to show better results in very large datasets like Imagenet where non-spiking CNN based architecture has shown tremendous results. Since the SN-CNN model developed integrates the spiking network and non-spiking network in NengoDL, the SN-CNN model if modified and tested in such dataset may provide better results owing to its better learning capability. Hence the performance analysis of SN-CNN model can be extended to various different datasets. Also, this research dealt with models involving 10 classes. It needs to be extended to more number of classes to test the robustness of this model. Hence, Cifar-100 datasets need to be tried and tested for improvement.

A further extension can be carried out if the feasibility of training using spiking neurons suffice. This will be of great advantage, as online training may extend the work to the development of unsupervised systems which is the need of the hour in big data.

7.2.3   Hardware Implementation of Bio-plausible Systems

This thesis presented the analysis of hardware implementation. However, actual implementation may result in identifying some more hardware related issues in design and hence it is necessary. The SN-CNN model was implemented in NengoDL which is a black box simulator and was recently developed as an open-source. From the architecture and the developers, it was found that the conversion of the developed model in NengoDL to FPGA platform is not available and is under development. Hence, once that was available the real implementation will give more insights into the design process. But, still, it can be made to run in neuromorphic hardware and can be analysed.

Visualisations were also limited in NengoDL and with more visualisation, better analysis is possible. However, some of the key ways of implementation in hardware include the opportunities available in a cloud-based environment and Tensor processing units. The SN-CNN model owing to its ease of implementation and ease of scaling can be extended to better and more complex classification systems and tested.

With lots of scope of improvement owing to the fast emerging research in this field, bio-plausible systems can in near future be attributed to fast detection and classification of images, similar to the way the human eye does. When this becomes a reality, the way computation systems work will transform into a different and new dimension shifting the entire focus from traditional Von Neumann systems to more bio-plausible systems.

Appendix 1

PYTHON SOURCE CODE FOR BASIC SN-CNN

```python
#matplotlib inline
from urllib.request import urlretrieve
import zipfile
import nengo
import nengo_dl
import tensorflow as tf
from tensorflow.examples.tutorials.mnist import input_data
import numpy as np
import matplotlib.pyplot as plt
from nengo.utils.matplotlib import rasterplot
```

1.1   INPUT PREPROCESSING OF FASHIONMNIST DATA

```python
mnist = input_data.read_data_sets("fashionMNIST data/", one_hot=True)
for i in range(6):
plt.figure()
plt.imshow(np.reshape(mnist.train.images[i], (28, 28)))
plt.axis('off')
plt.title(str(np.argmax(mnist.train.labels[i])));
```

1.1.1   Defining Parameters

```python
# lif parameters
```

```
lif_neurons = nengo.LIF(tau_rc=0.01, tau_ref=0.001)
# softlif parameters (lif parameters + sigma)
softlif_neurons = nengo_dl.SoftLIFRate(tau_rc=0.01, tau_ref=0.002,
sigma=0.001)
# ensemble parameters
ens_params = dict(max_rates=nengo.dists.Choice([100]),
intercepts=nengo.dists.Choice([0]))
```

1.1.2   Plot Some Example LIF Tuning Curves

```
for neuron_type in (lif_neurons, softlif_neurons):
with nengo.Network(seed=0) as net:
ens = nengo.Ensemble(10, 1, neuron_type=neuron_type)
with nengo_dl.Simulator(net) as sim:
plt.figure()
plt.plot(*nengo.utils.ensemble.tuning_curves(ens, sim))
plt.xlabel("input value")
plt.ylabel("firing rate")
plt.title(str(neuron_type))
```

1.2   BUILDING A CNN MODEL

```
def build_network(neuron_type):
with nengo.Network() as net:
# we'll make all the nengo objects in the network
# non-trainable. we could train them if we wanted, but they don't
# add any representational power so we can save some computation
# by ignoring them. note that this doesn't affect the internal
# components of tensornodes, which will always be trainable or
```

```python
# non-trainable depending on the code written in the tensornode.
nengo_dl.configure_settings(trainable=False)
# the input node that will be used to feed in input images
inp = nengo.Node(nengo.processes.PresentInput(mnist.test.images, 0.2))
# add the first convolutional layer
x = nengo_dl.tensor_layer( inp, tf.layers.conv2d, shape_in=(28, 28, 1),
filters=32, kernel_size=3)
# apply the neural nonlinearity
x = nengo_dl.tensor_layer(x, neuron_type, **ens_params)
# add another convolutional layer
x = nengo_dl.tensor_layer( x, tf.layers.conv2d, shape_in=(26, 26, 32),
filters=16, kernel_size=3)
x = nengo_dl.tensor_layer(x, neuron_type, **ens_params)
# apply the neural nonlinearity
x = nengo_dl.tensor_layer(x, neuron_type, **ens_params)
# add another convolutional layer
x = nengo_dl.tensor_layer( x, tf.layers.conv2d, shape_in=(24, 24, 16), filters=8,
kernel_size=3)
# add a pooling layer
x = nengo_dl.tensor_layer( x, tf.layers.max_pooling2d, shape_in=(22, 22, 8),
pool_size=2, strides=2)
# add a pooling layer
x = nengo_dl.tensor_layer( x, tf.layers.max_pooling2d, shape_in=(11, 11, 8),
pool_size=2, strides=2)
# add a dense layer, with neural nonlinearity.
# note that for all-to-all connections like this we can use the
# normal nengo connection transform to implement the weights
# (instead of using a separate tensor_layer).
x, conn = nengo_dl.tensor_layer( x, neuron_type, **ens_params,
transform=nengo_dl.dists.Glorot(), shape_in=(128,), return_conn=True)
```

```python
# we need to set the weights and biases to be trainable
# (since we set the default to be trainable=False)
# note: we used return_conn=True above so that we could access
# the connection object for this reason.
net.config[x].trainable = True
net.config[conn].trainable = True
# add a dropout layer
x = nengo_dl.tensor_layer(x, tf.layers.dropout, rate=0.3)
# the final 10 dimensional class output
x = nengo_dl.tensor_layer(x, tf.layers.dense, units=10)
return net, inp, x
# construct the network
net, inp, out = build_network(softlif_neurons)
with net:
out_p = nengo.Probe(out)
```

1.2.1   Construct the Simulator

```python
minibatch_size = 200
sim = nengo_dl.Simulator(net, minibatch_size=minibatch_size)
# note that we need to add the time dimension (axis 1), which has length 1
# in this case. we're also going to reduce the number of test images, just to
# speed up this example.
train_inputs = inp: mnist.train.images[:, None, :]
train_targets = out_p: mnist.train.labels[:, None, :]
test_inputs = inp: mnist.test.images[:minibatch_size*2, None, :]
test_targets = out_p: mnist.test.labels[:minibatch_size*2, None, :]
```

1.2.2   Defining Loss Function and Optimizer

```python
def objective(x, y):
```

```python
return tf.nn.softmax_cross_entropy_with_logits(logits=x, labels=y)
opt = tf.train.AdamOptimizer(learning_rate=0.001, beta1=0.9, beta2=0.999,
epsilon=1e-08,)
def classification_error(outputs, targets):
return 100 * tf.reduce_mean( tf.cast(tf.not_equal(tf.argmax(outputs[:, -1],
axis=-1),
tf.argmax(targets[:, -1], axis=-1)), tf.float32))
print("error before training: classification_error))
```

1.2.3   Training the Non-spiking Network and Saving the Weights

```python
do_training = True
if do_training:
# run training
sim.train(train_inputs, train_targets, opt, objective=objective, n_epochs=10)
# save the parameters to file
sim.save_params("./fashionmnist_paramsbs6")
else:
# load parameters
sim.load_params("./fashionmnist_paramsbs6")
print("error after training: classification_error))
sim.close()
```

1.3   RUNNING THE SPIKING NETWORK

```python
net, inp, out = build_network(lif_neurons)
with net:
out_p = nengo.Probe(out, synapse=0.1)
inp_p = nengo.Probe(inp)
```

```
sim = nengo_dl.Simulator(net, minibatch_size=minibatch_size,
unroll_simulation=10)
sim.load_params("./fashionmnist_paramsbs6")
n_steps = 80
test_inputs_time = inp: np.tile(v, (1, n_steps, 1)) for v in test_inputs.values()
test_targets_time = out_p: np.tile(v, (1, n_steps, 1)) for v in test_targets.values()
print("spiking neuron error: sim.run_steps(n_steps, input_feeds=inp:
test_inputs_time[inp][:minibatch_size])
```

1.3.1 Plotting Sample Test Images with Tuning Curves indicating their Classification

```
for i in range(5):
plt.figure()
plt.subplot(1, 2, 1)
plt.imshow(np.reshape(mnist.test.images[i], (28, 28)))
plt.axis('off')
plt.subplot(1, 2, 2)
plt.plot(sim.trange(), sim.data[out_p][i])
plt.legend([str(i) for i in range(10)], loc="upper left")
plt.xlabel("time")
sim.close()
```

REFERENCES

1. 'MS Windows NT kernel description', `https://www.kdnuggets.com/2016/07/softmax-regression-related-logistic-regression.html`, Accessed: 2018-05-22.

2. Abadi, M., Agarwal, A., Barham, P., Brevdo, E., Chen, Z., Citro, C., Corrado, G. S., Davis, A., Dean, J., Devin, M., et al. (2016a), 'Tensorflow: Large-scale machine learning on heterogeneous distributed systems', arXiv preprint arXiv:1603.04467.

3. Abadi, M., Barham, P., Chen, J., Chen, Z., Davis, A., Dean, J., Devin, M., Ghemawat, S., Irving, G., Isard, M., et al. (2016b), 'Tensorflow: A system for large-scale machine learning', 12th {USENIX} Symposium on Operating Systems Design and Implementation ({OSDI} 16), pp 265–283.

4. Abbas, S. S. A. and Muthulakshmi, C. (2014), 'Neuromorphic implementation of adaptive exponential integrate and fire neuron', Communication and Network Technologies (ICCNT), 2014 International Conference on, pp.233–237,IEEE.

5. Abbott, L. and Kepler, T. B. (1990), 'Model neurons: from hodgkin-huxley to hopfield', Statistical mechanics of neural networks, pp .5–18,Springer.

6. Abbott, L. F. (1999a), 'Lapicque's introduction of the integrate-and-fire model neuron (1907)', Brain research bulletin, Vol.50(no.5):pp.303–304.

7. Abbott, L. F. (1999b), 'Lapicque's introduction of the integrate-and-fire model neuron (1907)', Brain Research Bulletin, Vol.50(no.5-6):pp.303–304.

8. Abdelbaki, H., Gelenbe, E., and El-Khamy, S. E. (2000), 'Analog hardware implementation of the random neural network model', Neural Networks, 2000. IJCNN 2000, Proceedings of the IEEE-INNS-ENNS International Joint Conference on, volume 4, pp 197–201,IEEE.

9. Aibe, N., Yasunaga, M., Yoshihara, I., and Kim, J. H. (2002), 'A probabilistic neural network hardware system using a learning-parameter parallel architecture', Neural Networks, 2002. IJCNN'02. Proceedings of the 2002 International Joint Conference on, volume 3, pp 2270–2275,IEEE.

10. Akers, L., Walker, M., Ferry, D., and Grondin, R. (1989), 'A limited-interconnect, highly layered synthetic neural architecture', VLSI for artificial intelligence, pp 218–226,Springer.

11. Al-Rodhan, N. (2016(accessed August 21, 2017)), 'Neuromorphic Computers: What will they Change?'.

12. Ang, C. H., Jin, C., Leong, P. H., and van Schaik, A. (2011), 'Spiking neural network-based auto-associative memory using fpga interconnect delays', Field-Programmable Technology (FPT), 2011 International Conference on, pp 1–4,IEEE.

13. Arthur, J. V. and Boahen, K. A. (2011), 'Silicon-neuron design: A dynamical systems approach', IEEE Transactions on Circuits and Systems I: Regular Papers, Vol.58(no.5):p.1034–1043.

14. Aunet, S. and Hartmann, M. (2003), 'Real-time reconfigurable linear threshold elements and some applications to neural hardware', Evolvable Systems: From Biology to Hardware, pp pp.227–239.

15. Bañuelos-Saucedo, M., Castillo-Hernández, J., Quintana-Thierry, S., Damián-Zamacona, R., Valeriano-Assem, J., Cervantes, R., Fuentes-González, R., Calva-Olmos, G., and Pérez-Silva, J. (2003), 'Implementation of a neuron model using fpgas', Journal of applied research and technology, Vol.1(no.3):pp.248–255.

16. Basham, E. J. and Parent, D. W. (2009), 'An analog circuit implementation of a quadratic integrate and fire neuron', Engineering in Medicine and Biology Society, 2009. EMBC 2009. Annual International Conference of the IEEE, pp .741–744,IEEE.

17. Bekolay, T., Bergstra, J., Hunsberger, E., DeWolf, T., Stewart, T. C., Rasmussen, D., Choo, X., Voelker, A., and Eliasmith, C. (2014), 'Nengo: a python tool for building large-scale functional brain models', Frontiers in neuroinformatics, Vol.7:p.48.

18. Benjamin, B. V., Gao, P., McQuinn, E., Choudhary, S., Chandrasekaran, A. R., Bussat, J.-M., Alvarez-Icaza, R., Arthur, J. V., Merolla, P. A., and Boahen, K. (2014), 'Neurogrid: A mixed-analog-digital multichip system for large-scale neural simulations', Proceedings of the IEEE, Vol.102(no.5):pp.699–716.

19. Berzish, M., Eliasmith, C., and Tripp, B. (2016), 'Real-time fpga simulation of surrogate models of large spiking networks', International Conference on Artificial Neural Networks, pp 349–356,Springer.

20. Bezborah, A. (2012), 'A hardware architecture for training of artificial neural networks using particle swarm optimization', Intelligent Systems, Modelling and Simulation (ISMS), 2012 Third International Conference on, pp 67–70,IEEE.

21. Bishop, C., Bishop, C. M., et al. (1995), 'Neural networks for pattern recognition', Oxford university press.

22. Bobier, B., Stewart, T. C., and Eliasmith, C. (2014), 'A unifying mechanistic model of selective attention in spiking neurons', PLoS computational biology, Vol.10(no.6):e1003577.

23. Bofill, A., Thompson, D., and Murray, A. F. (2002), 'Citcuits for vlsi implementation of temporally asymmetric hebbian learning', Advances in Neural Information processing systems, pp 1091–1098.

24. Brink, S., Koziol, S., Ramakrishnan, S., and Hasler, P. (2008), 'A biophysically based dendrite model using programmable floating-gate devices', Circuits and Systems, 2008. ISCAS 2008. IEEE International Symposium on, pp 432–435,IEEE.

25. Brink, S., Nease, S., and Hasler, P. (2013), 'Computing with networks of spiking neurons on a biophysically motivated floating-gate based neuromorphic integrated circuit', Neural Networks,Vol.45:pp.39–49.

26. Buhry, L., Saïghi, S., Salem, W. B., and Renaud, S. (2009), 'Adjusting neuron models in neuromimetic ics using the differential evolution algorithm', Neural Engineering, 2009. NER'09. 4th International IEEE/EMBS Conference on, pp 681–684,IEEE.

27. Burr, G. W., Shelby, R. M., Sebastian, A., Kim, S., Kim, S., Sidler, S., Virwani, K., Ishii, M., Narayanan, P., Fumarola, A., et al. (2017), 'Neuromorphic computing using non-volatile memory', Advances in Physics: X, Vol.2(no.1):pp.89–124.

28. Burr, G. W., Shelby, R. M., Sidler, S., Di Nolfo, C., Jang, J., Boybat, I., Shenoy, R. S., Narayanan, P., Virwani, K., Giacometti, E. U., et al. (2015), 'Experimental demonstration and tolerancing of a large-scale neural network (165 000 synapses) using phase-change memory as the synaptic weight element', IEEE Transactions on Electron Devices, Vol.62(no.11):pp.3498–3507.

29. Burr, J. B. (1991), 'Digital neural network implementations', Neural networks, concepts, applications, and implementations, Vol.3:pp.237–285.

30. Caelli, T., Guan, L., and Wen, W. (1999), 'Modularity in neural computing', Proceedings of the IEEE, Vol.87(no.9):pp.1497–1518.

31. Calimera, A., Macii, E., and Poncino, M. (2013), 'The human brain project and neuromorphic computing', Functional neurology, Vol.28(no.3):p191.

32. Cauwenberghs, G. (1994), 'A learning analog neural network chip with continuous-time recurrent dynamics', Advances in Neural Information Processing Systems, pp 858–865.

33. Choo, F.-X. and Eliasmith, C. (2010), 'A spiking neuron model of serial-order recall', Proceedings of the Annual Meeting of the Cognitive Science Society, volume 32.

34. Collins, D. R. and Penz, P. A. (1989), 'Considerations for neural network hardware implementations', Circuits and Systems, 1989., IEEE International Symposium on, pp 834–836,IEEE.

35. Collobert, R., Kavukcuoglu, K., and Farabet, C. (2011), 'Torch7: A matlab-like environment for machine learning', Technical report.

36. Cong, J. and Xiao, B. (2011), 'mrfpga: A novel fpga architecture with memristor-based reconfiguration', 2011 IEEE/ACM international symposium on Nanoscale Architectures, pp 1–8, IEEE.

37. Davison, A. P., Brüderle, D., Eppler, J. M., Kremkow, J., Muller, E., Pecevski, D., Perrinet, L., and Yger, P. (2009), 'Pynn: a common interface for neuronal network simulators', Frontiers in neuroinformatics, Vol.2:no.11.

38. Deco, G., Jirsa, V. K., Robinson, P. A., Breakspear, M., and Friston, K. (2008), 'The dynamic brain: from spiking neurons to neural masses and cortical fields', PLoS computational biology, Vol.4(no.8):e1000092.

39. Delorme, A., Gautrais, J., Van Rullen, R., and Thorpe, S. (1999), 'Spikenet: A simulator for modeling large networks of integrate and fire neurons', Neurocomputing, Vol.26:pp.989–996.

40. Deshmukh, A., Morghade, J., Khera, A., and Bajaj, P. (2005), 'Binary neural networks–a cmos design approach', Knowledge-Based Intelligent Information and Engineering Systems, pp 153–153, Springer.

41. DeWolf, T., Stewart, T. C., Slotine, J.-J., and Eliasmith, C. (2016), 'A spiking neural model of adaptive arm control', Proceedings of the Royal Society B: Biological Sciences, Vol.283(no.1843):p.20162134.

42. Diehl, P. U., Zarrella, G., Cassidy, A., Pedroni, B. U., and Neftci, E. (2016), 'Conversion of artificial recurrent neural networks to spiking neural networks for low-power neuromorphic hardware', Rebooting Computing (ICRC), IEEE International Conference on, pp 1–8, IEEE.

43. Donati, E., Corradi, F., Stefanini, C., and Indiveri, G. (2014), 'A spiking implementation of the lamprey's central pattern generator in neuromorphic vlsi', Biomedical Circuits and Systems Conference (BioCAS), 2014 IEEE, pp 512–515, IEEE.

44. Eliasmith, C., Stewart, T. C., Choo, X., Bekolay, T., DeWolf, T., Tang, Y., and Rasmussen, D. (2012), 'A large-scale model of the functioning brain', science, Vol.338(no.6111):pp.1202–1205.

45. Farquhar, E., Gordon, C., and Hasler, P. (2006), 'A field programmable neural array', Circuits and Systems, 2006. ISCAS 2006. Proceedings. 2006 IEEE International Symposium on, pp 4–pp, IEEE.

46. Fieres, J., Schemmel, J., and Meier, K. (2006), 'Training convolutional networks of threshold neurons suited for low-power hardware implementation', Neural Networks, 2006. IJCNN'06. International Joint Conference on, pp 21–28, IEEE.

47. Folowosele, F., Hamilton, T. J., and Etienne-Cummings, R. (2011), 'Silicon modeling of the mihalas,–niebur neuron', IEEE transactions on neural networks, Vol.22(no.12):pp.1915–1927.

48. Freeh, V. W., Pan, F., Kappiah, N., Lowenthal, D. K., and Springer, R. (2005), 'Exploring the energy-time tradeoff in mpi programs on a power-scalable cluster', Parallel and Distributed Processing Symposium, 2005. Proceedings. 19th IEEE International, pp 10–pp,IEEE.

49. Fujita, O. and Amemiya, Y. (1993), 'A floating-gate analog memory device for neural networks', IEEE Transactions on Electron Devices, Vol.40(no.11):pp.2029–2035.

50. Fukushima, K. (1980), 'Neocognitron: A self-organizing neural network for a mechanism of pattern recognition unaffected by shift in position', Biological Cybernetics, Vol.36(no.4):pp.193–202.

51. Furber, S. (2016), 'Large-scale neuromorphic computing systems', Journal of neural engineering, Vol.13(no.5):p.051001.

52. Gaines, B. R. (1969), 'Stochastic computing systems', Advances in information systems science, pp 37–172,Springer.

53. Gewaltig, M.-O. and Diesmann, M. (2007), 'Nest (neural simulation tool)', Scholarpedia, Vol.2(no.4):p.1430.

54. Ghani, A., McGinnity, T. M., Maguire, L. P., and Harkin, J. (2006), 'Area efficient architecture for large scale implementation of biologically plausible spiking neural networks on reconfigurable hardware', Field Programmable Logic and Applications, 2006. FPL'06. International Conference on, pp 1–2,IEEE.

55. Glackin, B., McGinnity, T. M., Maguire, L. P., Wu, Q., and Belatreche, A. (2005), 'A novel approach for the implementation of large scale spiking neural networks on fpga hardware', International Work-Conference on Artificial Neural Networks, pp 552–563,Springer.

56. Graf, H. and De Vegvar, P. (1987), 'A cmos associative memory chip based on neural networks', Solid-State Circuits Conference. Digest of Technical Papers. 1987 IEEE International, volume 30, pp 304–305,IEEE.

57. Grassia, F., Levi, T., Kohno, T., and Saïghi, S. (2014), 'Silicon neuron: digital hardware implementation of the quartic model', Artificial Life and Robotics, Vol.19(no.3):pp.215–219.

58. Han, B., Sengupta, A., and Roy, K. (2016), 'On the energy benefits of spiking deep neural networks: A case study', Neural Networks (IJCNN), 2016 International Joint Conference on, pp 971–976,IEEE.

59. Harkin, J., Morgan, F., McDaid, L., Hall, S., McGinley, B., and Cawley, S. (2009), 'A reconfigurable and biologically inspired paradigm for computation using network-on-chip and spiking neural networks', International Journal of Reconfigurable Computing, Vol.2009:p.2.

60. Hasler, P. and Akers, L. (1990), 'Vlsi neural systems and circuits', Computers and Communications, 1990. Conference Proceedings., Ninth Annual International Phoenix Conference on, pp 31–37,IEEE.

61. Hayati, M., Nouri, M., Haghiri, S., and Abbott, D. (2016), 'A digital realization of astrocyte and neural glial interactions', IEEE transactions on biomedical circuits and systems, Vol.10(no.2):pp.518–529.

62. Haykin, S. and Network, N. (2004), 'A comprehensive foundation', Neural Networks, Vol.2(2004):p.41.

63. Heemskerk, J. (1995), 'Overview of neural hardware. neurocomputers for brain-style processing. design, implementation and application', PhD Thesis, Unit of Experimental and Theoretical Psychology Leiden University.

64. Henderson, R. (2014 (accessed August 21, 2017)), 'Intel claims that by 2026 processors will have as many transistors as there are neurons in a brain'.

65. Hikawa, H. (2003), 'A digital hardware pulse-mode neuron with piecewise linear activation function', IEEE Transactions on Neural Networks, Vol.14(no.5):pp.1028–1037.

66. Hindmarsh, J. and Cornelius, P. (2005), 'The development of the hindmarsh-rose model for bursting', Bursting: the genesis of rhythm in the nervous system, pp .3–18,World Scientific.

67. Hines, M. L. and Carnevale, N. T. (1997), 'The neuron simulation environment', Neural computation, Vol.9(no.6):pp.1179–1209.

68. Hishiki, T. and Torikai, H. (2009), 'Bifurcation analysis of a resonate-and-fire-type digital spiking neuron', Neural Information Processing, pp .392–400,Springer.

69. Hishiki, T. and Torikai, H. (2010), 'Neural behaviors and nonlinear dynamics of a rotate-and-fire digital spiking neuron', Neural Networks (IJCNN), The 2010 International Joint Conference on, pp .1–8,IEEE.

70. Hollis, P. W. and Paulos, J. J. (1992), 'An analog bicmos hopfield neuron', Analog Integrated Circuits and Signal Processing, Vol.2(no.4):pp.273–279.

71. Hsieh, H.-Y. and Tang, K.-T. (2013), 'Hardware friendly probabilistic spiking neural network with long-term and short-term plasticity', IEEE transactions on neural networks and learning systems, Vol.24(no.12):pp.2063–2074.

72. Hsu, J. (2014), 'Ibm's new brain [news]', IEEE spectrum, Vol.51(no.10):pp.17–19.

73. Hubel, D. H. and Wiesel, T. (1962), 'Receptive fields, binocular interaction, and functional architecture in the cat's visual cortex', Journal of Physiology (London), Vol.160:pp.106–154.

74. Hunsberger, E. and Eliasmith, C. (2015), 'Spiking deep networks with lif neurons', arXiv preprint arXiv:1510.08829.

75. Hylander, P., Meader, J., and Frie, E. (1993), 'Vlsi implementation of pulse coded winner take all networks', Circuits and Systems, 1993., Proceedings of the 36th Midwest Symposium on, pp 758–761,IEEE.

76. Hynna, K. M. and Boahen, K. (2006), 'Neuronal ion-channel dynamics in silicon', Circuits and Systems, 2006. ISCAS 2006. Proceedings. 2006 IEEE International Symposium on, p .4,IEEE.

77. Ibrayev, T., James, A. P., Merkel, C., and Kudithipudi, D. (2016), 'A design of htm spatial pooler for face recognition using memristor-cmos hybrid circuits', Circuits and Systems (ISCAS), 2016 IEEE International Symposium on, pp 1254–1257,IEEE.

78. Isobe, K. and Torikai, H. (2016), 'A novel hardware-efficient asynchronous cellular automaton model of spike-timing-dependent synaptic plasticity', IEEE Transactions on Circuits and Systems II: Express Briefs, Vol.63(no.6):pp.603–607.

79. Ivakhnenko, A. G. (1968), 'The group method of data handling – a rival of the method of stochastic approximation', Soviet Automatic Control, Vol.13(no.3):pp.43–55.

80. Ivakhnenko, A. G. (1971), 'Polynomial theory of complex systems', IEEE transactions on Systems, Man, and Cybernetics, Vol.1(no.4):pp.364–378.

81. Izhikevich, E. M. (2003), 'Simple model of spiking neurons', IEEE Transactions on neural networks, Vol.14(no.6):pp.1569–1572.

82. Izhikevich, E. M. and FitzHugh, R. (2006), 'Fitzhugh-nagumo model', Scholarpedia, Vol.1(no.9):p.1349.

83. Jang, J.-W., Attarimashalkoubeh, B., Prakash, A., Hwang, H., and Jeong, Y.-H. (2016), 'Scalable neuron circuit using conductive-bridge ram for pattern reconstructions', IEEE Transactions on Electron Devices, Vol.63(no.6):pp.2610–2613.

84. Jia, Y., Shelhamer, E., Donahue, J., Karayev, S., Long, J., Girshick, R., Guadarrama, S., and Darrell, T. (2014), 'Caffe: Convolutional architecture for fast feature embedding', Proceedings of the 22nd ACM international conference on Multimedia, pp 675–678,ACM.

85. Johnston, S., Prasad, G., Maguire, L., and McGinnity, M. (2005), 'Comparative investigation into classical and spiking neuron implementations on fpgas', International Conference on Artificial Neural Networks, pp 269–274,Springer.

86. Kamavisdar, P., Saluja, S., and Agrawal, S. (2013), 'A survey on image classification approaches and techniques', International Journal of Advanced Research in Computer and Communication Engineering, Vol.2(no.1):pp.1005–1009.

87. Kanazawa, Y., Asai, T., and Amemiya, Y. (2003), 'A hardware depressing synapse and its application to contrast-invariant pattern recognition', SICE 2003 Annual Conference, volume 2, pp 1558–1563,IEEE.

88. Kay, K. N. (2017), 'Principles for models of neural information processing', Neuroimage.

89. Kay, K. N. (2018), 'Principles for models of neural information processing', Neuroimage, Vol.180:pp.101–109.

90. Khan, M. M., Lester, D. R., Plana, L. A., Rast, A., Jin, X., Painkras, E., and Furber, S. B. (2008), 'Spinnaker: mapping neural networks onto a massively-parallel chip multiprocessor', Neural Networks, 2008. IJCNN 2008.(IEEE World Congress on Computational Intelligence). IEEE International Joint Conference on, pp 2849–2856,IEEE.

91. Kingma, D. P. and Ba, J. (2014), 'Adam: A method for stochastic optimization', arXiv preprint arXiv:1412.6980.

92. Knight, B. W. (1972), 'Dynamics of Encoding in a Population of Neurons', The Journal of General Physiology, Vol.59(no.6):pp.734–766.

93. Knight, J., Voelker, A. R., Mundy, A., Eliasmith, C., and Furber, S. (2016), 'Efficient spinnaker simulation of a heteroassociative memory using the neural engineering framework', Neural Networks (IJCNN), 2016 International Joint Conference on, pp 5210–5217,IEEE.

94. Koch, C. (1999), 'Biophysics of computation: Information processing in single neurons', Oxford University Press, New York,NY.

95. Krid, M., Dammak, A., and Masmoudi, D. S. (2006), 'Fpga implementation of programmable pulse mode neural network with on chip learning for signature application', Electronics, Circuits and Systems, 2006. ICECS'06. 13th IEEE International Conference on, pp 942–945,IEEE.

96. Kriegeskorte, N. (2015), 'Deep neural networks: a new framework for modeling biological vision and brain information processing', Annual review of vision science, Vol.1:pp. 417–446.

97. Kröger, B. J., Bekolay, T., and Eliasmith, C. (2014), 'Modeling speech production using the neural engineering framework', Cognitive Infocommunications (CogInfoCom), 2014 5th IEEE Conference on, pp 203–208,IEEE.

98. Kudithipudi, D., Saleh, Q., Merkel, C., Thesing, J., and Wysocki, B. (2015), 'Design and analysis of a neuromemristive reservoir computing architecture for biosignal processing', Frontiers in neuroscience, Vol.9.

99. Kuo, Y.-H., Kao, C.-I., and Chen, J.-J. (1993), 'A fuzzy neural network model and its hardware implementation', IEEE Transactions on Fuzzy Systems, Vol.1(no.3):pp.171–183.

100. Lapicque, L. (1907), 'Recherches quantitatifs sur l'excitation electrique des nerfs traitée comme une polarisation.', J. Physiol. Paris, Vol.9:pp.620–635.

101. Lecar, H. (2007), 'Morris-lecar model', Scholarpedia, Vol.2(no.10):p.1333.

102. LeCun, Y., Bengio, Y., and Hinton, G. (2015), 'Deep learning', nature, Vol.521(no.7553):p.436.

103. LeCun, Y., Boser, B. E., Denker, J. S., Henderson, D., Howard, R. E., Hubbard, W. E., and Jackel, L. D. (1990), 'Handwritten digit recognition with a back-propagation network', Advances in neural information processing systems, pp 396–404.

104. LeCun, Y., Bottou, L., Bengio, Y., and Haffner, P. (1998), 'Gradient-based learning applied to document recognition', Proc. IEEE, Vol.86(no.11):pp.2278–2324.

105. Leong, P. H. and Jabri, M. A. (1992), 'A vlsi neural network for morphology classification', Neural Networks, 1992. IJCNN., International Joint Conference on, volume 2, pp 678–683,IEEE.

106. Liao, Y. (2001), 'Neural networks in hardware: A survey', Department of Computer Science, University of California.

107. Liu, M., Yu, H., and Wang, W. (2008), 'Fpaa based on integration of cmos and nanojunction devices for neuromorphic applications', International Conference on Nano-Networks, pp 44–48,Springer.

108. Lu, C., Hong, C., and Chen, H. (2007), 'A scalable and programmable architecture for the continuous restricted boltzmann machine in vlsi', Circuits and Systems, 2007. ISCAS 2007. IEEE International Symposium on, pp 1297–1300,IEEE.

109. Lu, D. and Weng, Q. (2007), 'A survey of image classification methods and techniques for improving classification performance', International Journal of Remote Sensing, Vol.28(no.5):pp.823–870.

110. Maass, W. (1997), 'Networks of spiking neurons: the third generation of neural network models', Neural networks, Vol.10(no.9):pp.1659–1671.

111. Maeda, Y. and Tada, T. (2003), 'Fpga implementation of a pulse density neural network with learning ability using simultaneous perturbation', IEEE Transactions on Neural Networks, Vol.14(no.3):pp.688–695.

112. Mainen, Z. F. and Sejnowski, T. J. (1995), 'Reliability of spike timing in neocortical neurons.', Science, Vol.268(no.5216):pp.1503–6.

113. Mann, J. R. and Gilbert, S. (1989), 'An analog self-organizing neural network chip', Advances in neural information processing systems, pp 739–747.

114. Matsubara, T. and Torikai, H. (2011), 'Dynamic response behaviors of a generalized asynchronous digital spiking neuron model', Neural Information Processing, pp .395–404,Springer.

115. McCulloch, W. S. and Pitts, W. (1943), 'A logical calculus of the ideas immanent in nervous activity', The bulletin of mathematical biophysics, Vol.5(no.4):pp.115–133.

116. McGinnity, T., Roche, B., Maguire, L., and McDaid, L. (1998), 'Novel architecture and synapse design for hardware implementations of neural networks', Computers & electrical engineering, Vol.24(no.1):pp.75–87.

117. Merkel, C., Kudithipudi, D., and Sereni, N. (2013), 'Periodic activation functions in memristor-based analog neural networks', Neural Networks (IJCNN), The 2013 International Joint Conference on, pp 1–7,IEEE.

118. Mosa, A. H., Kyamakya, K., Ali, M., and Al, F., 'Neuro-computing based matrix inversion concept involving cellular neural networks black-box training concept'.

119. Muthuramalingam, A., Himavathi, S., and Srinivasan, E. (2008), 'Neural network implementation using fpga: issues and application', International journal of information technology, Vol.4(no.2):pp.86–92.

120. Nair, M. V. and Dudek, P. (2015), 'Gradient-descent-based learning in memristive crossbar arrays', Neural Networks (IJCNN), 2015 International Joint Conference on, pp 1–7,IEEE.

121. NirmalaDevi, M., Mohankumar, N., and Arumugam, S. (2009), 'Modeling and analysis of neuro–genetic hybrid system on fpga', Elektronika ir Elektrotechnika, Vol.96(no.8):pp.43–46.

122. Noack, M., Krause, M., Mayr, C., Partzsch, J., and Schuffny, R. (2014), 'Vlsi implementation of a conductance-based multi-synapse using switched-capacitor circuits', Circuits and Systems (ISCAS), 2014 IEEE International Symposium on, pp 850–853,IEEE.

123. Oster, M., Wang, Y., Douglas, R., and Liu, S.-C. (2008), 'Quantification of a spike-based winner-take-all vlsi network', IEEE Transactions on Circuits and Systems I: Regular Papers, Vol.55(no.10):pp.3160–3169.

124. Pande, S., Morgan, F., Cawley, S., Bruintjes, T., Smit, G., McGinley, B., Carrillo, S., Harkin, J., and McDaid, L. (2013), 'Modular neural tile architecture for compact embedded hardware spiking neural network', Neural processing letters, Vol.38(no.2):pp.131–153.

125. Papert, S. A. (1966), 'The summer vision project'.

126. Paugam-Moisy, H. (2006), 'Spiking neuron networks a survey', Technical report, IDIAP.

127. Paugam-Moisy, H. and Bohte, S. (2012), 'Computing with spiking neuron networks', Handbook of natural computing, pp 335–376,Springer.

128. Peper, F. (2017), 'The end of moore's law: Opportunities for natural computing?', New Generation Computing, Vol.35(no.3):pp.253–269.

129. Pérez-Carrasco, J. A., Zhao, B., Serrano, C., Acha, B., Serrano-Gotarredona, T., Chen, S., and Linares-Barranco, B. (2013), 'Mapping from frame-driven to frame-free event-driven vision systems by low-rate rate coding and coincidence processing–application to feedforward convnets', IEEE transactions on pattern analysis and machine intelligence, Vol.35(no.11):pp.2706–2719.

130. Polsky, A., Mel, B. W., and Schiller, J. (2004), 'Computational subunits in thin dendrites of pyramidal cells', Nature Neuroscience, Vol.7(no.6):pp.621–627.

131. Ponulak, F. and Kasinski, A. (2011), 'Introduction to spiking neural networks: Information processing, learning and applications.', Acta neurobiologiae experimentalis, Vol.71(no.4):pp.409–433.

132. Poon, C.-S. and Zhou, K. (2011), 'Neuromorphic silicon neurons and large-scale neural networks: challenges and opportunities', Frontiers in neuroscience, Vol.5.

133. Rachmuth, G. and Poon, C.-S. (2003), 'Design of a neuromorphic hebbian synapse using analog vlsi', Neural Engineering, 2003. Conference Proceedings. First International IEEE EMBS Conference on, pp 221–224,IEEE.

134. Ranjan, J. A. K., Sigamani, T., and Barnabas, J. (2019), 'A novel and efficient classifier using spiking neural network', The Journal of Supercomputing, pp pp.1–16.

135. Rasmussen, D. (2018), 'Nengodl: Combining deep learning and neuromorphic modelling methods', arXiv preprint arXiv:1805.11144.

136. Rasmussen, D. and Eliasmith, C. (2014), 'A spiking neural model applied to the study of human performance and cognitive decline on raven's advanced progressive matrices', Intelligence, Vol.42:pp.53–82.

137. Ruder, S. (2016), 'An overview of gradient descent optimization algorithms', arXiv preprint arXiv:1609.04747.

138. Rueckauer, B., Lungu, I.-A., Hu, Y., Pfeiffer, M., and Liu, S.-C. (2017), 'Conversion of continuous-valued deep networks to efficient event-driven networks for image classification', Frontiers in neuroscience, Vol.11:p.682.

139. Rumelhart, D. E., Hinton, G. E., Williams, R. J., et al. (1988), 'Learning representations by back-propagating errors', Cognitive modeling, Vol.5(no.3):p.1.

140. Sahin, I. and Koyuncu, I. (2012), 'Design and implementation of neural networks neurons with radbas, logsig, and tansig activation functions on fpga', Elektronika ir Elektrotechnika, Vol.120(no.4):pp.51–54.

141. Salam, F. M. (1989), 'A model of neural circuits for programmable vlsi implementation', Circuits and Systems, 1989., IEEE International Symposium on, pp 849–851,IEEE.

142. Schneider, C. and Card, H. (1991), 'Cmos implementation of analog hebbian synaptic learning circuits', Neural Networks, 1991., IJCNN-91-Seattle International Joint Conference on, volume 1, pp 437–442,IEEE.

143. Schoenauer, T., Jahnke, A., Roth, U., and Klar, H. (1998), 'Digital neurohardware: principles and perspectives', Proceedings of Neuronal Networks in Applications '98, pp pp.101–106.

144. Schuman, C. D., Potok, T. E., Patton, R. M., Birdwell, J. D., Dean, M. E., Rose, G. S., and Plank, J. S. (2017), 'A survey of neuromorphic computing and neural networks in hardware', arXiv preprint arXiv:1705.06963.

145. Sengupta, A., Ye, Y., Wang, R., Liu, C., and Roy, K. (2018), 'Going deeper in spiking neural networks: Vgg and residual architectures', arXiv preprint arXiv:1802.02627.

146. Shayani, H. (2013), 'A Practical Investigation into Achieving Bio-Plausibility in Evo-Devo Neural Microcircuits Feasible in an FPGA', PhD thesis, UCL (University College London).

147. Sheri, A. M., Rafique, A., Pedrycz, W., and Jeon, M. (2015), 'Contrastive divergence for memristor-based restricted boltzmann machine', Engineering Applications of Artificial Intelligence, Vol.37:pp.336–342.

148. Shi, Y., Nguyen, L., Oh, S., Liu, X., Koushan, F., Jameson, J. R., and Kuzum, D. (2018), 'Neuroinspired unsupervised learning and pruning with subquantum cbram arrays', Nature communications, Vol.9(no.1):p.5312.

149. Skrbek, M. (1999), 'Fast neural network implementation', Neural Network World, Vol.9(no.5):pp.375–391.

150. Snider, G. S. (2008), 'Spike-timing-dependent learning in memristive nanodevices', Proceedings of the 2008 IEEE International Symposium on Nanoscale Architectures, pp 85–92,IEEE Computer Society.

151. Srinivasan, V., Dugger, J., and Hasler, P. (2005), 'An adaptive analog synapse circuit that implements the least-mean-square learning rule', Circuits and Systems, 2005. ISCAS 2005. IEEE International Symposium on, pp 4441–4444,IEEE.

152. Stewart, T. C., Bekolay, T., and Eliasmith, C. (2012), 'Learning to select actions with spiking neurons in the basal ganglia', Frontiers in neuroscience, Vol.6:p.2.

153. Stewart, T. C., Choo, X., Eliasmith, C., et al. (2010), 'Dynamic behaviour of a spiking model of action selection in the basal ganglia', Proceedings of the 10th international conference on cognitive modeling, pp 235–40,Citeseer.

154. Stewart, T. C. and Eliasmith, C. (2009), 'Spiking neurons and central executive control: The origin of the 50-millisecond cognitive cycle', 9th International Conference on Cognitive Modelling, volume 122, pp 130–131,University of Manchester Manchester.

155. Stimberg, M., Goodman, D. F., Benichoux, V., and Brette, R. (2013), 'Brian 2-the second coming: spiking neural network simulation in python with code generation', BMC neuroscience, Vol.14(no.1):p.38.

156. Stromatias, E., Neil, D., Pfeiffer, M., Galluppi, F., Furber, S. B., and Liu, S.-C. (2015), 'Robustness of spiking deep belief networks to noise and reduced bit precision of neuro-inspired hardware platforms', Frontiers in neuroscience, Vol.9.

157. Strukov, D. B., Snider, G. S., Stewart, D. R., and Williams, R. S. (2008), 'The missing memristor found', nature, Vol.453(no.7191):pp.80.

158. Strukov, D. B., Stewart, D. R., Borghetti, J., Li, X., Pickett, M., Ribeiro, G. M., Robinett, W., Snider, G., Strachan, J. P., Wu, W., et al. (2010), 'Hybrid cmos/memristor circuits', Proceedings of 2010 IEEE International Symposium on Circuits and Systems, pp 1967–1970,IEEE.

159. Szabó, T., Antoni, L., Horváth, G., and Fehér, B. (2000), 'A full-parallel digital implementation for pre-trained nns', Neural Networks, 2000. IJCNN 2000, Proceedings of the IEEE-INNS-ENNS International Joint Conference on, volume 2, pp 49–54,IEEE.

160. Tal, D. and Schwartz, E. L. (2006), 'Computing with the leaky integrate-and-fire neuron: logarithmic computation and multiplication', Computing, Vol.9(no.2).

161. Tamukoh, H. and Sekine, M. (2010), 'A dynamically reconfigurable platform for self-organizing neural network hardware', Neural Information Processing. Models and Applications, pp pp.439–446.

162. Team, T. T. D., Al-Rfou, R., Alain, G., Almahairi, A., Angermueller, C., Bahdanau, D., Ballas, N., Bastien, F., Bayer, J., Belikov, A., et al. (2016), 'Theano: A python framework for fast computation of mathematical expressions', arXiv preprint arXiv:1605.02688.

163. Thorpe, S., Fize, D., and Marlot, C. (1996), 'Speed of processing in the human visual system', nature, 381(6582):520.

164. Tuma, T., Le Gallo, M., Sebastian, A., and Eleftheriou, E. (2016), 'Detecting correlations using phase-change neurons and synapses', IEEE Electron Device Letters, Vol.37(no.9):pp.1238–1241.

165. Ueda, M., Nishitani, Y., Kaneko, Y., and Omote, A. (2014), 'Back-propagation operation for analog neural network hardware with synapse components having hysteresis characteristics', PloS one, Vol.9(no.11):e112659.

166. VanRullen, R., Guyonneau, R., and Thorpe, S. J. (2005), 'Spike times make sense', Trends in neurosciences, Vol.28(no.1):pp.1–4.

167. Vianello, E., Garbin, D., Bichler, O., Piccolboni, G., Molas, G., De Salvo, B., and Perniola, L. (2017), 'Multiple binary oxrams as synapses for convolutional neural networks', Advances in Neuromorphic Hardware Exploiting Emerging Nanoscale Devices, pp 109–127,Springer.

168. Walker, M. and Akers, L. (1988), 'A neuromorphic approach to adaptive digital circuitry', Computers and Communications, 1988. Conference Proceedings., Seventh Annual International Phoenix Conference on, pp 19–23,IEEE.

169. Wang, Z., Ma, Y., Cheng, F., and Yang, L. (2010), 'Review of pulse-coupled neural networks', Image and Vision Computing, Vol.28(no.1):pp.5–13.

170. Weinfeld, M. (1989), '6.2 a fully digital integrated cmos hopfield network including the learning algorithm', THE KLUWER INTERNATIONAL SERIES IN ENGINEERING AND COMPUTER SCIENCE VLSI, COMPUTER ARCHITECTURE AND DIGITAL SIGNAL PROCESSING, p p.169.

171. Xiao, H., Rasul, K., and Vollgraf, R. (2017), 'Fashion-mnist: a novel image dataset for benchmarking machine learning algorithms', arXiv preprint arXiv:1708.07747.

172. Yamakawa, T. (1996), 'Silicon implementation of a fuzzy neuron', IEEE transactions on fuzzy systems, Vol.4(no.4):pp.488–501.

173. Yang, J., Ahmadi, M., Jullien, G. A., and Miller, W. C. (1999), 'An in-the-loop training method for vlsi neural networks', Circuits and Systems, 1999. ISCAS'99. Proceedings of the 1999 IEEE International Symposium on, volume 5, pp 619–622,IEEE.

174. Yang, J. J., Pickett, M. D., Li, X., Ohlberg, D. A., Stewart, D. R., and Williams, R. S. (2008), 'Memristive switching mechanism for metal/oxide/metal nanodevices', Nature nanotechnology, Vol.3(no.7):pp.429.

175. Yao, E., Hussain, S., Basu, A., and Huang, G.-B. (2013), 'Computation using mismatch: Neuromorphic extreme learning machines', Biomedical Circuits and Systems Conference (BioCAS), 2013 IEEE, pp 294–297,IEEE.

176. Yuille, A. L. and Grzywacz, N. M. (1989), 'A winner-take-all mechanism based on presynaptic inhibition feedback', Neural Computation, Vol.1(no.3):pp.334–347.

177. Zhu, J. and Sutton, P. (2003), 'Fpga implementations of neural networks–a survey of a decade of progress', International Conference on Field Programmable Logic and Applications, pp 1062–1066,Springer.

178. Zhu, X., Shen, J., Chi, B., and Wang, Z. (2005), 'Circuit implementation of multi-thresholded neuron (mtn) using bicmos technology', Neural Networks, 2005. IJCNN'05. Proceedings. 2005 IEEE International Joint Conference on, volume 1, pp 627–632,IEEE.

179. Zomaya, A. Y. (2006), 'Handbook of nature-inspired and innovative computing: integrating classical models with emerging technologies', Springer Science & Business Media.